Understanding Modern Mathematics

Saul Stahl, PhD
University of Kansas

Paul E. Johnson, PhD
University of Kansas

JONES AND BARTLETT PUBLISHERS
Sudbury, Massachusetts
BOSTON   TORONTO   LONDON   SINGAPORE

World Headquarters
Jones and Bartlett Publishers
40 Tall Pine Drive
Sudbury, MA 01776
978-443-5000
info@jbpub.com
www.jbpub.com

Jones and Bartlett Publishers Canada
6339 Ormindale Way
Mississauga, Ontario L5V 1J2
CANADA

Jones and Bartlett Publishers International
Barb House, Barb Mews
London W6 7PA
UK

Jones and Bartlett's books and products are available through most bookstores and online booksellers. To contact Jones and Bartlett Publishers directly, call 800-832-0034, fax 978-443-8000, or visit our website www.jbpub.com.

Substantial discounts on bulk quantities of Jones and Bartlett's publications are available to corporations, professional associations, and other qualified organizations. For details and specific discount information, contact the special sales department at Jones and Bartlett via the above contact information or send an email to specialsales@jbpub.com.

6048
Production Credits
Acquisitions Editor: Tim Anderson
Production Director: Amy Rose
Editorial Assistant: Kate Koch
Production Editor: Tracey Chapman
Marketing Manager: Andrea DeFronzo
Manufacturing Buyer: Therese Connell

Composition: Publishers' Design and
    Production Services, Inc.
Cover Design: Timothy Dziewit
Cover Image: © Phil Kestell/Shuttershock, Inc.
Printing and Binding: Malloy, Incorporated
Cover Printing: Malloy, Incorporated

Library of Congress Cataloging-in-Publication Data
Stahl, Saul.
    Understanding modern mathematics / Saul Stahl, Paul Johnson. — 1st ed.
        p. cm.
    Includes bibliographical references and index.
    ISBN-13: 978-0-7637-3401-5 (pbk.)
    1. Mathematics.   I. Johnson, Paul E.   II. Title.
    QA37.3.S725 2006
    510—dc22

                                                        2005030297

Printed in the United States of America
10 09 08 07 06   10 9 8 7 6 5 4 3 2 1

The authors are indebted to their colleagues James Church and Ben Cobb for several informative discussions regarding probability and statistics. They also acknowledge several helpful comments by Mark Huneke, Jennifer Wagner, and several anonymous reviewers.

Credits

Chapter 3 is a condensation of portions of the first author's *A Gentle Introduction to Game Theory*, American Mathematical Society, Mathematical World Volume 13, Copyright © 1991, Providence RI. The authors are grateful to the society for their generous permission to include these selections here.

The games described in Exercises 4.2.12–4.2.16 and 4.5.41–4.5.44 are reprinted from J. D. Williams's *The Compleat Strategyst*, Dover Publishing Co., NY, 1982 with the permission of the RAND Corporation.

Contents

Preface — vii

Chapter 1 **Probability** — 1
1.1 The Quantification of Likelihood — 1
1.2 Galileo Galilei (A Historical Interlude) — 10
1.3 Binomial Processes — 15
1.4 The Normal Approximation — 22
1.5 Activities/Websites — 31

Chapter 2 **Statistics** — 33
2.1 Summarizing Data — 33
2.2 Totaling Measurement Errors — 43
2.3 Least Squares and Regression — 50
2.4 Normally Distributed Data — 58
2.5 Statistical Inference — 71
2.6 The Central Limit Theorem — 82
2.7 Activities/Websites — 86

Chapter 3 **Voting Systems** — 87
3.1 Introduction — 87
3.2 Mathematical Concepts — 88
3.3 The Plurality Problem — 89
3.4 Rank-Order Voting: The Borda Count — 96
3.5 Sequential Pairwise Comparisons — 107
3.6 Condorcet Methods: The Round Robin Tournament — 113
3.7 Single Vote Systems: Cousins of Plurality and
Majority Rule — 130
3.8 Conclusion — 135

Chapter 4 **Game Theory** — 137
4.1 Introduction to Games — 137
4.2 Formal Zero-Sum Games — 148
4.3 Pure Strategies for Nonzero-Sum Games — 157
4.4 Mixed Strategies for Nonzero-Sum Games — 169

4.5   The Solution of Zero-Sum Games   188
4.6   Activities/Websites   204

Chapter 5   **Linear Programming**   **207**
5.1   Graphing Inequalities   207
5.2   Optimization   222

Chapters 6 and 7 are available online at the publisher's website.

Appendix   **A-1**

Table A
A Table of the Normal Distribution   A-1

Table B
A Table of Binomial Probabilities   A-3

References and Further Readings   **R-1**

Solutions to Selected Exercises   **S-1**

Index   **I-1**

Preface

This book is aimed at the popular "Topics in Mathematics" course that is a requirement for the B.A. degree in many universities. This course has no prerequisites above and beyond a minimal amount of high school algebra and is meant to better acquaint the students with mathematics through an exposure to its applications and an exposition of some of its culture.

All the textbooks written for this course have adopted the survey approach and touch many topics lightly. We, however, believe that the course's goals can be best accomplished by focusing on a few important topics and explaining both their significance and evolution in some depth. This calls for locating good mathematical ideas that are presentable at the level of the class, convincing the students of their value, and allowing enough time to permit the concepts to take root in the students' minds through drill and development.

In our opinion, it is a serious pedagogical mistake to teach easy applications just because they show how mathematics can be used in an applied context. The superficiality of the mathematics will not escape the students, and they will lose respect for the subject matter. With the respect will also go their willingness to learn. For this reason we have concentrated on a few topics whose centrality to mathematics is well known: probability, statistics, game theory, linear programming, and symmetry. The topic of social choice was added to these because its paradoxes and surprises make it a favorite with students.

Although these topics have been expounded in many texts, our development of most of them is unique:

Probability and statistics: The exposition is focused on the normal curve. The evolution of this curve is set forth in detail together with explanations of the scientific needs that motivated the invention of some of the standard statistical procedures. This amounts to a comprehensible explanation of how and why the normal curve came to occupy the eminent position it does in our society—a tale in which Kepler, Galileo, the French Revolution, Gauss, and the Average Man all play their parts.

Voting systems: This chapter explores voting mechanisms that can be used to decide among a discrete set of alternatives. Although individual voters may be transitive and logical in their behavior, the aggregation of their opinions through a voting procedure can often produce problematic outcomes. The chapter includes a wide range of voting procedures, including the long-standing ones like plurality, majority, and rank-order voting as well as some very recent proposals intended to address the weaknesses of the older methods. Several criteria for evaluation of voting procedures are considered.

Game theory: Enough game theory is taught so that the students gain an appreciation of the Nash equilibrium, a concept for which the economists voted to award the 1994 Nobel Prize to the mathematician John Nash. The popularity of the movie *A Beautiful Mind* makes this subject even more topical. Nontrivial applications to economics, population biology, anthropology, political science, and other social situations are discussed

Linear programming: The exposition is conventional. We included this material because of its popularity among the instructors of this course and because of its importance as a tool of applied mathematics.

The following topic is available online at the publisher's website.

http://math.jbpub.com/catalog/0763734012/

Symmetry: Contrary to common practice in books for this audience, we added three-dimensional symmetry to the standard two-dimensional material. This makes possible visual/calculational problems that have just the right level of difficulty. It also provides a context for recounting to them the mysterious tale of Monstrous Moonshine—the story of the classification of the finite simple groups, MONSTER, and the mysterious connections with non-Euclidean geometry.

Chapter 1
Probability

Animal astragali—bones used for gambling—have been found in graves that are thousands of years old. These archaeologic artifacts indicate that the interest in the likelihood of events goes back to antiquity. Probability, which is the quantification of the informal notion of likelihood, dates back only as far as the Italian Renaissance. Subjective and objective notions of probability are discussed, and some formal properties of the latter are stated. These are applied to the derivation of the probabilities of compound events. The normal curve is introduced as a computational aid.

1.1   The Quantification of Likelihood

It is often said that the only certain things in life are death and taxes. Most other aspects of life and the universe have some uncertainty attached to them: Although the sun is practically certain to rise tomorrow, it is known that some day it will fail to do so; on the next July Fourth the temperature in Kansas City may top 85°F; the Dow Jones index may rise tomorrow; a woman may be elected as president in the next U.S. presidential election. All readers will agree that these events have been listed in decreasing order of likelihood, and this consensus indicates that the ephemeral quality of likelihood should in fact be quantifiable. Such quantifications were first phrased in terms of one's willingness to risk stakes in betting on such uncertain events, that is, as odds. This, however, was too subjective a procedure for scientific purposes, and eventually an objective set of rules were laid down that serve as an objective definition of probability.

The quantification of uncertainty is sometimes easy and sometimes hard. When a coin is flipped, a probability of 1/2 is assigned to the eventuality that it will come up *heads*. The coin might, of course, be weighted so as to render this inaccurate, but we think of this possibility as an irrelevant aberration. If the coin is fair, as most coins are believed to be, then the probability of *heads* is 1/2, and only a mathematician might quibble. But how can the likelihood of a woman being elected president be quantified objectively?

In these pages it is first shown how to assign probabilities in elementary, almost trivial, situations. These simple contexts are then combined so as to cover more complex and realistic questions.

A *random experiment* is a future incident with a set of possible *outcomes*. The set of outcomes is called the *sample space*. Thus, the flip of a coin is a random experiment with two outcomes, *heads* and *tails*, and the sample space {*heads, tails*}. The toss of a standard die is a random experiment with six outcomes: 1, 2, 3, 4, 5, or 6. The associated sample space is {1, 2, 3, 4, 5, 6}. The selection of a shirt to wear is also a random experiment whose sample space consists of all the shirts in one's wardrobe.

Depending on one's interests, different sample spaces can be associated with the same random experiment. Suppose a pair of dice is tossed. If we are interested in the sum of the faces, then the sample space is {2, 3, 4, 5, 6, 7, 8, 9, 10, 11, 12}. If, on the other hand, we are interested in the number of ones, then the sample space is {0, 1, 2}.

Given a random experiment and a sample space, each of the outcomes has a probability that quantifies the likelihood of that outcome transpiring when the experiment is performed. This probability is not defined explicitly but instead is computed, wherever possible, on the basis of some elementary properties that are universally accepted. If a random experiment has a finite number of outcomes, then each of the outcomes has a probability between 0 and 1, and these probabilities are so distributed that

P1.   *The sum of the probabilities of all the outcomes is 1;*

P2.   *Equally likely outcomes have equal probabilities.*

If a coin is flipped, then each of the outcomes *heads* and *tails* has probability 1/2. If a die is tossed, and there is no reason to doubt its fairness, then each of the outcomes 1, 2, 3, 4, 5, or 6 has probability 1/6. In general, if a random experiment has n outcomes and there is no reason to believe that any of these outcomes are more likely to take place than others, then each of the outcomes is assigned the probability

$$\frac{1}{n}$$

Properties P1 and P2 above are themselves the consequences of a deeper assumption, which can *almost* be taken as a definition, that the probability of an outcome in some sense describes its frequency if the experiment is performed a large number of times. It is tempting to say that the probability of a flipped coin coming up *heads* is 1/2 because when a coin is tossed, say, a thousand times, it will come up *heads* approximately 500 times. The reason this cannot be taken as a genuine definition is that the same coin will also come up *heads* approximately 499 times, and so the very same repeated experiment could be used to justify assigning a probability of 0.499 to *heads*. A new experiment for deciding between 0.499 and 0.5 as the appropriate probability can of course be designed, but, among other difficulties, this would call for millions of flips, and it has certainly not been carried out by either the author or any of his acquaintances. Nevertheless, we are all willing to recognize 0.5 as the correct probability on logical grounds alone. The reason for this is that although the actual frequencies fail to serve as a definition of the probabilities, the expected (but rarely realized) frequencies do have the following analogs of the properties 1 and 2 above:

F1.   *The actual frequencies of all the outcomes add up to the total number of iterations of the experiment;*

F2.   *The frequencies of equally likely outcomes are expected to be equal.*

Properties F1 and F2 of frequencies then translate into properties P1 and P2 of probabilities.

When the outcomes are not equally likely, some reasoning is needed to compute their probabilities. One method for doing this is to "break down" outcomes into "simpler" outcomes that are equally likely. Such a "complex" outcome is called an *event*. More formally, an event is a set of outcomes. For example, if a die is tossed, one may speak of the event that its outcome is *odd*, which is identified with the set of outcomes {1, 3, 5}. Similarly, the event that the outcome is greater than 4 can be thought of as the set {5, 6}. The following property can now be added to P1 and P2:

P3.   *The probability of an event is the sum of the probabilities of its constituent outcomes.*

When all the outcomes of the random experiment in question are equally likely, this is equivalent to

P3'. $Probability\ of\ event = \dfrac{number\ of\ constituent\ outcomes}{total\ number\ of\ outcomes}$

Hence, the probability of a tossed die coming up even is

$$\frac{3}{6} = \frac{1}{2}$$

and the probability of its coming up greater than 4 is

$$\frac{2}{6} = \frac{1}{3}$$

Suppose two dice are tossed. The 36 equally likely outcomes are then

$$
\begin{array}{cccccc}
(1, 1) & (1, 2) & (1, 3) & (1, 4) & (1, 5) & (1, 6) \\
(2, 1) & (2, 2) & (2, 3) & (2, 4) & (2, 5) & (2, 6) \\
(3, 1) & (3, 2) & (3, 3) & (3, 4) & (3, 5) & (3, 6) \\
(4, 1) & (4, 2) & (4, 3) & (4, 4) & (4, 5) & (4, 6) \\
(5, 1) & (5, 2) & (5, 3) & (5, 4) & (5, 5) & (5, 6) \\
(6, 1) & (6, 2) & (6, 3) & (6, 4) & (6, 5) & (6, 6)
\end{array}
$$

and each of these outcomes has probability $\dfrac{1}{36}$. The event that the two dice show the same face is the set

$$\{(1, 1), (2, 2), (3, 3), (4, 4), (5, 5), (6, 6)\}$$

and so has probability

$$\frac{6}{36} = \frac{1}{6}$$

On the other hand, the event that one of the faces is more than double the other is the set

$$\{(1, 3), (1, 4), (1, 5), (1, 6), (2, 5), (2, 6), (3, 1), (4, 1), (5, 1), (5, 2), (6, 1), (6, 2)\}$$

and has probability

$$\frac{12}{36} = \frac{1}{3}$$

As an example of a random experiment whose outcomes are not equally likely, we gave the addition of the faces that show after a pair of dice is tossed. The sample space is $\{2, 3, 4, 5, 6, 7, 8, 9, 10, 11, 12\}$, and the probability of each of the outcomes is computed by expressing it as a set of the outcomes of the previous experiment:

Sum	The event as a set	Probability
2	$\{(1, 1)\}$	$\frac{1}{36}$
3	$\{(1, 2), (2, 1)\}$	$\frac{2}{36} = \frac{1}{18}$
4	$\{(1, 3), (2, 2), (3, 1)\}$	$\frac{3}{36} = \frac{1}{12}$
5	$\{(1, 4), (2, 3), (3, 2), (4, 1)\}$	$\frac{4}{36} = \frac{1}{9}$
6	$\{(1, 5), (2, 4), (3, 3), (4, 2), (5, 1)\}$	$\frac{5}{36}$
7	$\{(1, 6), (2, 5), (3, 4), (4, 3), (5, 2), (6, 1)\}$	$\frac{6}{36} = \frac{1}{6}$
8	$\{(2, 6), (3, 5), (4, 4), (5, 3), (6, 2)\}$	$\frac{5}{36}$
9	$\{(3, 6), (4, 5), (5, 4), (6, 3)\}$	$\frac{4}{36} = \frac{1}{9}$
10	$\{(4, 6), (5, 5), (6, 4)\}$	$\frac{3}{36} = \frac{1}{12}$
11	$\{(5, 6), (6, 5)\}$	$\frac{2}{36} = \frac{1}{18}$
12	$\{(6, 6)\}$	$\frac{1}{36}$

Two events are said to be *complementary* if between them they constitute the entire sample space of a random experiment *and* they share no outcomes. Such are *heads* and *tails* for the coin flip and *even* and *odd* for the toss of a die. Not getting a *two* upon tossing a die is complementary to getting it.

P4.   *If A and B are complementary events, then*

$$\text{Probability of A} + \text{Probability of B} = 1$$

Another common method for expressing the outcomes of one experiment as a combination of outcomes of another experiment is that of conjunction. Outcome A is said to be a *conjunction* of B and C if its occurrence is logically equivalent to the occurrence of both B and C. Such, for example, is the case for the following events when two dice are thrown:

A: The sum of the faces showing is 12;

B: The first die shows a 6;

C: The second die shows a 6.

Similarly, if two coins are flipped, then outcome A below is the conjunction of B and C:

A: No heads are showing;

B: The first coin came up *tails*;

C: The second coin came up *tails*.

Two random experiments are said to be *independent* if information about the eventual outcome of either sheds no light on the eventual outcome of the other. If two coins are flipped, then knowing that the first coin came up *heads* tells us nothing about what happens to the second coin. These two coin flips are therefore independent. Similarly, the tossing of two dice are also independent experiments. On the other hand, if two coins are glued *heads* to *heads* and flipped, then knowing that the first coin landed tails down tells us that the second coin landed *tails* up. The two coin tosses are therefore not independent of each other. Similarly, the height and the weight of a randomly selected person are both random experiments. However, because tall people tend to be heavier than short ones, these are not independent experiments.

Suppose now that event A is the conjunction of events B and C that are respectively associated with independent random experiments. Then

P5.   *Probability of A = (Probability of B) · (Probability of C)*

Accordingly, if two dice are tossed, then the probability of the first and the second coming up 5 and 6, respectively, is

$$\frac{1}{6} \cdot \frac{1}{6} = \frac{1}{36}$$

Similarly, if a coin is flipped and a die is tossed, then the probability that the coin comes up *heads* and the die shows a 3 is

$$\frac{1}{2} \cdot \frac{1}{6} = \frac{1}{12}$$

The multiplication principle P5 is not limited to just two experiments. The probability of a conjunction of events associated with any number of pairwise independent experiments is the product of the probabilities of the individual events. Thus, if a family with five children is selected at random, then the probability of all of them being daughters is

$$\frac{1}{2} \cdot \frac{1}{2} \cdot \frac{1}{2} \cdot \frac{1}{2} \cdot \frac{1}{2} = \frac{1}{32}$$

Similarly, if a die is tossed four times, then the probability of its showing 1, 2, 3, and 4 in succession is

$$\frac{1}{6} \cdot \frac{1}{6} \cdot \frac{1}{6} \cdot \frac{1}{6} = \frac{1}{1296}$$

EXERCISES 1.1

1. Two standard dice are tossed. Compute the probabilities of the following events:

    a) The two faces showing differ by 1.

    b) The two faces showing differ by no more than 1.

    c) The two faces showing differ by 2.

    d) The two faces showing differ by no more than 2.

    e) The two faces showing are both even.

    f) The two faces showing are both odd.

    g) Exactly one of the faces showing is a 5.

    h) At least one of the faces showing is a 5.

    i) At most one of the faces showing is a 5.

2. If the eight triangular faces of the solid of Figure 1.1 are marked with 1, 2, ..., 8, respectively, then it can be used as a fair eight-sided die. When two such eight-sided dice are tossed, the sample space is

(1, 1) (1, 2) (1, 3) (1, 4) (1, 5) (1, 6) (1, 7) (1, 8)
(2, 1) (2, 2) (2, 3) (2, 4) (2, 5) (2, 6) (2, 7) (2, 8)
(3, 1) (3, 2) (3, 3) (3, 4) (3, 5) (3, 6) (3, 7) (3, 8)
(4, 1) (4, 2) (4, 3) (4, 4) (4, 5) (4, 6) (4, 7) (4, 8)
(5, 1) (5, 2) (5, 3) (5, 4) (5, 5) (5, 6) (5, 7) (5, 8)
(6, 1) (6, 2) (6, 3) (6, 4) (6, 5) (6, 6) (6, 7) (6, 8)
(7, 1) (7, 2) (7, 3) (7, 4) (7, 5) (7, 6) (7, 7) (7, 8)
(8, 1) (8, 2) (8, 3) (8, 4) (8, 5) (8, 6) (8, 7) (8, 8)

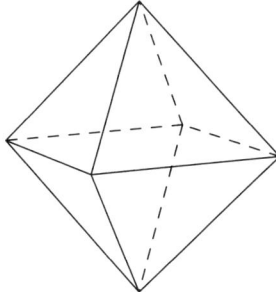

Figure 1.1   *The octahedron, an eight-sided die.*

Compute the probabilities of the following events when two 8-sided dice are tossed:

a)   The two faces showing differ by 1.

b)   The two faces showing differ by no more than 1.

c)   The two faces showing differ by 2.

d)   The two faces showing differ by no more than 2.

e)   The two faces showing are both even.

f)   The two faces showing are both odd.

g)   Exactly one of the faces showing is a 5.

h)   At least one of the faces showing is a 5.

i)   At most one of the faces showing is a 5.

3.   Two four-sided dice (Fig. 1.2) are tossed. Write down the sample space of this experiment.

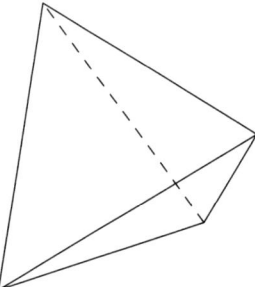

Figure 1.2   *The tetrahedron, a four-sided die.*

4. Two four-sided dice are tossed. Compute the probabilities of the following events:

a) The two faces showing differ by 1.

b) The two faces showing differ by no more than 1.

c) The two faces showing differ by 2.

d) The two faces showing differ by no more than 2.

e) The two faces showing are both even.

f) The two faces showing are both odd.

g) Exactly one of the faces showing is a 3.

h) At least one of the faces showing is a 3.

i) At most one of the faces showing is a 3.

5. A standard six-sided die and a four-sided die are tossed. Write down the sample space of this random experiment. Compute the probabilities of the following events:

a) The two faces showing differ by 1.

b) The two faces showing differ by no more than 1.

c) The two faces showing differ by 2.

d) The two faces showing differ by no more than 2.

e) The two faces showing are both even.

f) The two faces showing are both odd.

g) Exactly one of the faces showing is a 3.

h) At least one of the faces showing is a 3.

i) At most one of the faces showing is a 3.

6. Two four-sided dice (Fig. 1.2) are tossed and the faces showing are added up. What is the corresponding sample space and what is the probability of each outcome?

7. Two 12-sided dice (Fig. 1.3) are tossed. Compute the probabilities of the following events:

   a) The two faces showing differ by 1.

   b) The two faces showing differ by no more than 1.

   c) The two faces showing differ by 2.

   d) The two faces showing differ by no more than 2.

   e) The two faces showing are both even.

   f) The two faces showing are both odd.

   g) Exactly one of the faces showing is a 3.

   h) At least one of the faces showing is a 3.

   i) At most one of the faces showing is a 3.

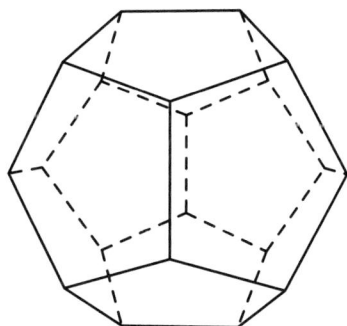

Figure 1.3   *The dodecahedron, a 12-sided die.*

8. Two eight-sided dice are thrown. Compute the probability that the two faces showing add up to the following:

   a) 3   b) 4   c) 5   d) 6   e) 7   f) 8   g) 9

9. Two 12-sided dice are thrown. Compute the probability that the two faces showing add up to the following:

   a) 3   b) 4   c) 5   d) 6   e) 7   f) 8   g) 9

   h) 10   i) 11   j) 12   k) 13   l) 14   m) 15   n) 16

10. A card is selected at random from a standard deck of 52. Compute the probabilities of the following events:

   a) The card is a diamond.

   b) The card is red.

   c) The card is a 4.

   d) The card is a face card (i.e., jack, queen, or king).

   e) The card is a number between 5 and 7 (inclusive).

   f) The card is a red ace.

11. Flip a coin on a table top and cover it with your hand before you see the outcome. Explain why the probability of seeing *heads* when the coin is uncovered is *not* 0.5. Can you evaluate this probability? If you had to bet on the outcome, what odds would be acceptable to you?

1.2   Galileo Galilei (A Historical Interlude)

Around the year 1620 Galileo Galilei (1564–1643), by his own account, was ordered to consider the following question regarding the relative likelihoods of certain sums occurring when three dice are thrown:

The numbers 9, 10, 11, and 12 can each be expressed as the sum of the faces of three dice in six ways, namely,

$$9: \; 6+2+1 = 5+3+1 = 5+2+2 = 4+4+1 = 4+3+2 = 3+3+3$$

$$10: \; 6+3+1 = 6+2+2 = 5+4+1 = 5+3+2 = 4+4+2 = 4+3+3$$

$$11: \; 6+4+1 = 6+3+2 = 5+5+1 = 5+4+2 = 5+3+3 = 4+4+3$$

$$12: \; 6+5+1 = 6+4+2 = 6+3+3 = 5+5+2 = 5+4+3 = 4+4+4$$

Why then does the experience of dice players indicate that 10 and 11 are more advantageous than 9 and 12?

By that time Galileo was one of the best known scientists in Europe, and it is hard to imagine anyone *ordering* him to do anything except for his employer, the Grand Duke of Tuscany. In his response, Galileo explained to the Duke that the advantage (what is now called the likelihood or probability) of a number should be measured not by the frequency with which this number can be made up as a sum, but by the number of actual throws that realize that number as their sum.

To clarify this distinction, the sums on the previous page are referred to as *3-partitions* or simply *partitions*. It is clear from Galileo's letter that the Duke, assuming he was indeed the interlocutor, believed that the frequency of a number's 3-partitions should be the measure of its advantage. Not so, said Galileo, for this ignored the fact that 3-partitions can be produced in different ways. Of course, the partition $3 = 1 + 1 + 1$ can only be produced by the single throw of three ones, and the same goes for the partition $6 = 2 + 2 + 2$. However, the partition $4 = 2 + 1 + 1$ can be produced by three different throws: when the first die shows 2 and the second and third show the 1, or the second die a 2 and the first and third the 1, or the third die a 2 and the first and second the 1. This observation is summarized as

Partition:	$2 + 1 + 1$
	2, 1, 1
Throws:	1, 2, 1
	1, 1, 2

Similarly, the partition $8 = 3 + 3 + 2$ can also be produced by three different throws:

Partition:	$3 + 3 + 2$
	2, 3, 3
Throws:	3, 2, 3
	3, 3, 2

On the other hand, if a partition's three summands are all different, then it can be produced in six ways. For example, for $8 = 4 + 3 + 1$:

Partition:	$4 + 3 + 1$
	1, 3, 4
	1, 4, 3
Throws:	3, 1, 4
	4, 1, 3
	3, 4, 1
	4, 3, 1

Taking these frequencies into account, the number of throws that produce the sums of 9 and 10, respectively, is shown as follows:

Sum = 9		Sum = 10	
Partition	No. of throws	Partition	No. of throws
$6 + 2 + 1$	6	$6 + 3 + 1$	6
$5 + 3 + 1$	6	$6 + 2 + 2$	3
$5 + 2 + 2$	3	$5 + 4 + 1$	6
$4 + 4 + 1$	3	$5 + 3 + 2$	6
$4 + 3 + 2$	6	$4 + 4 + 2$	3
$3 + 3 + 3$	1	$4 + 3 + 3$	3
Total:	25	Total:	27

Thus, there are 25 different throws that produce a sum of 9 and 27 that produce a 10. This difference, contended Galileo, accounted for the gamblers' aforementioned observation that 10 is more advantageous than 9. Having produced this explanation, Galileo also derived the number of throws that yield all the other possible sums and commented that the advantages of all the other configurations of the three-dice game could be measured by these frequencies.

It is interesting that no explicit probabilities were computed in this letter, which is one of the earliest documents wherein the likelihoods of unpredictable events are discussed in a mathematical manner. Galileo did compute the total number of distinct throws:

> . . . since a die has six faces, and when thrown it can equally well fall on any one of these, only 6 throws can be made with it, each different from all the others. But if together with the first die we throw a second, which also has six faces, we can make 36 throws each different from all the others, since each face of the first die can be combined with each of the second, and in consequence can make 6 different throws, whence it is clear that such combinations are 6 times 6, i.e., 36. And if we add a third die, since each of its six faces can be combined with each of the 36 combinations of the other two dice, we shall find that the combinations of three dice are 6 times 36, i.e., 216, each different from the others.

This done, however, he did not do what would be natural to us, which is to divide 25 by 216 and say that

$$\frac{25}{216}$$

is the probability of 9 occurring as the sum of a throw of three dice. He only used the number 216 to check by adding that the frequencies derived for the various partitions did indeed account for all possible outcomes. The notion of the probability of an event was only implicit in his thoughts. The closest he came to such a concept in this context was in the term "advantage." Elsewhere (see Section 2.2) he also used "frequency" to denote likelihood. This bears testimony to the excruciating slowness with which fundamental mathematical concepts that seem natural to us actually evolved. Further examples of such resistance to new ideas are given later.

It is interesting that the Duke's request was not for the solution of a practical problem but for a mathematical explanation of an observed phenomenon. His experience had already told him that 10 is more advantageous than 9, and he just wanted to know why this was so. In other words, he was motivated by the same belief that drove the classical Greek culture as well as the Renaissance era of which he was a prime mover: the belief that *Man is the Measure* and that the human mind is capable of imposing order on a seemingly inchoate Nature. This belief, that knowledge is worth having for its own sake and that the exercise of our logical faculties is as beneficial to our mental well-being as physical exercise is to our bodies, is still a part of our own ethos today, and it is in this spirit that the following variations on the Duke's games are offered here.

It was remarked above that the octahedron of Figure 1.1 can be used as a fair eight-sided die, and Galileo's analysis is now applied to the computation of the probability of the sum of the faces of a throw of three octahedral dice adding up to 9. The relevant partitions are

$$9 = 7 + 1 + 1 = 6 + 2 + 1 = 5 + 3 + 1 = 5 + 2 + 2 = 4 + 4 + 1 = 4 + 3 + 2 = 3 + 3 + 3$$

Rather than list the throws that give rise to each partition we merely record their number:

Partition:	7 + 1 + 1	6 + 2 + 1	5 + 3 + 1	5 + 2 + 2
No. of throws:	3	6	6	3
	4 + 4 + 1	4 + 3 + 2	3 + 3 + 3	
	3	6	1	

for a total of $3 + 6 + 6 + 3 + 3 + 6 + 1 = 28$. Because the total number of possible throws is $8 \cdot 8 \cdot 8 = 512$, it follows that the requisite probability is

$$\frac{28}{512} = \frac{7}{128} \approx 5.5\%$$

Other types of fair dice abound (see Figs. 1.1 to 1.3 and Exercises 15–17). The method for the computation of the probability of any number n occurring as the sum of the faces of three d-faced dice is now clear:

a) List the 3-partitions of n into three summands, none of which exceed d. (Note that once you have stipulated that the first die shows some integer, say k, then the faces of the second and third must add up to $n - k$. Keep in mind that none of the faces exceeds d and none exceeds the previous one.)

b) Replace each partition by its corresponding count of 6 or 3 or 1.

c) Add all the counts of part b and divide the sum by d^3.

Historical Note. Galileo made no claim to originality in his letter, and it can be inferred from the lack of argumentation in this document that these concepts were well understood by the experts of the time. In fact, the Italian mathematician Gerolamo Cardano (1501–1576) had already written his *Book on Games of Chance* more than 50 years earlier that contained, among many others, an analysis of the very same three-dice game discussed above. However, as this book was not published until 1663, Galileo was probably unaware of its existence. No other books on probability are known to have been written by that time, and so it is possible that Galileo was simply basing his analysis on what he considered to be common sense.

EXERCISES 1.2

1. Three six-sided dice are thrown. Compute the probability that the three faces showing add up to the following:

    a) 3 b) 4 c) 5 d) 6 e) 7 f) 8 g) 11

    h) 12 i) 13 j) 14 k) 15 l) 16 m) 17 n) 18

2. Three eight-sided dice are thrown (Fig. 1.1). Compute the probability that the three faces showing add up to the following:

    a) 3 b) 4 c) 5 d) 6 e) 7 f) 8 g) 9

    h) 10 i) 11 j) 12 k) 13 l) 14 m) 15 n) 16

3. Three four-sided dice are thrown (Fig. 1.2). Compute the probability that the three faces showing add up to the following:

    a) 3 b) 4 c) 5 d) 6 e) 7 f) 8 g) 9

4. Three 12-sided dice are thrown (see Fig. 1.3). Compute the probability that the three faces showing add up to the following:

    a) 3 b) 4 c) 5 d) 6 e) 7 f) 8 g) 9

    h) 10 i) 11 j) 12 k) 13 l) 14 m) 15 n) 16

5. List all the 4-partitions of the following integers:

    a) 4 b) 5 c) 6 d) 7 e) 8 f) 9 g) 10

6. List all the 24 different throws that correspond to a 4-partition, all of whose summands are distinct. (Suggestion: work with the partition $4 + 3 + 2 + 1$.)

7. List all the 12 different throws that correspond to a 4-partition, two of whose summands are identical. (Suggestion: work with the partition $3 + 3 + 2 + 1$.)

8. List all the six different throws that correspond to a 4-partition that has two pairs of identical summands. (Suggestion: work with the partition $2 + 2 + 1 + 1$.)

9. List all the four different throws that correspond to a 4-partition that has three identical summands. (Suggestion: work with the partition $2 + 1 + 1 + 1$.)

10. Explain why there is only one throw that corresponds to a 4-partition, all of whose summands are identical.

In solving Exercises 11–16 you may make use of Exercises 6–10.

11. Four four-sided dice are thrown. Compute the probability that the four faces showing add up to the following:

    a) 4 b) 5 c) 6 d) 7 e) 8 f) 9 g) 10

12. Four six-sided dice are thrown. Compute the probability that the four faces showing add up to the following:

    a) 4   b) 5   c) 6   d) 7   e) 8   f) 9   g) 10

13. Four eight-sided dice are thrown. Compute the probability that the four faces showing add up to the following:

    a) 4   b) 5   c) 6   d) 7   e) 8   f) 9   g) 10

14. Four 12-sided dice are thrown. Compute the probability that the four faces showing add up to the following:

    a) 4   b) 5   c) 6   d) 7   e) 8   f) 9   g) 10

15. Design a fair die whose faces are numbered 1–10.

16. Design a fair die whose faces are numbered 1–5. (Hint: have each number appear twice.)

17. Design a fair die whose faces are numbered 1–7.

1.3   Binomial Processes

The three-dice toss described in the previous section is actually a relatively complicated random phenomenon and does not constitute the traditional starting point for the formal study of probability. There are after all 16 possible outcomes (3 through 18), each of which can be broken down into several possible tosses. The *simplest* genuinely random event has only two possible outcomes and is called a *binomial*, or *Bernoulli*, *trial*, after the Swiss mathematical pioneer James Bernoulli (1654–1705) whose text, *Ars Conjectandi*, did much to establish the study of probabilities as a bona fide mathematical discipline. The flip of a coin, a baseball game, and the sex of an offspring are all examples of such bivalued binomial trials. Although the person performing the binomial trial may have no preference for one outcome over the other, it is nevertheless customary to designate one of the outcomes as *success* and the other as *failure*. This nomenclature is not meant as a moral judgment of the two—its purpose is merely to provide us with words for discussing the situation. Probabilists could just as well have chosen to designate the two alternatives in the more neutral, though less colorful, terms of *outcome I* and *outcome II*. Happily, they did not. The decision as to which outcome should be chosen as the *success* is unimportant—all that is needed is some designation.

  Given such a binomial trial, the probability of success, that is, the probability that the outcome labeled *success* actually occurs is denoted by p. The probability of failure is then $1 - p$. For example, suppose a coin is weighted so that upon being flipped it will come up *heads* 6 times out of 10. If *heads* is considered to be the success, then $p = 0.6$; if *tails* is the successful outcome, then $p = 0.4$. In the case of a fair coin, $p = 0.5$ regardless of which outcome is deemed to be a success. If 15% of all cars of a certain model are defective to some extent, and one such car is selected at random and examined for defects, then $p = 0.15$. Note that here success is tantamount to selecting a defective car, an occurrence that most of us would consider regrettable, to say the least. Such are the vagaries of formal mathematical terminology. Anyone unhappy with this "misappellation" could of course designate the selection of a non-defective car as success, which would entail $p = 0.85$.

A *binomial process* consists of several repetitions of identical, though independent, binomial trials. Ten successive flips of a coin are an example of such a process, as are the births of several offspring. The iterations need not follow each other in time; they may be performed simultaneously. Such is the case when three dice are tossed or a handful of coins is thrown. It is reasonable when performing such a set of trials to ask for the likelihood of a certain number of successes: If a coin is flipped 100 times, what is the probability of its coming up *heads* 50 times? If a randomly selected family has three children, what is the probability of two of them being girls?

There is a formula for these probabilities. Specifically, if a process consists of n independent repetitions of the same binomial trial with success probability p, then the probability of exactly k successes is

$$b(k; n, p) = \binom{n}{k} p^k (1 - p)^{n-k}$$

where $\binom{n}{k}$ (pronounced *n choose k*) is called *the choice number* and is to be computed by the formula

$$\binom{n}{k} = \frac{n(n-1)(n-2)\ldots(n-k+1)}{1 \cdot 2 \cdot 3 \cdot \ldots \cdot k}$$

The computation of the denominator of this expression is straightforward. The numerator is more easily understood if one realizes that it contains exactly as many factors as the denominator. Thus,

$$\binom{3}{2} = \frac{3 \cdot 2}{1 \cdot 2} = 3 \qquad \binom{4}{2} = \frac{4 \cdot 3}{1 \cdot 2} = 6 \qquad \binom{4}{3} = \frac{4 \cdot 3 \cdot 2}{1 \cdot 2 \cdot 3} = 4$$

$$\binom{5}{4} = \frac{5 \cdot 4 \cdot 3 \cdot 2}{1 \cdot 2 \cdot 3 \cdot 4} = 5 \qquad \binom{5}{3} = \frac{5 \cdot 4 \cdot 3}{1 \cdot 2 \cdot 3} = 10 \qquad \binom{5}{2} = \frac{5 \cdot 4}{1 \cdot 2} = 10$$

$$\binom{7}{4} = \frac{7 \cdot 6 \cdot 5 \cdot 4}{1 \cdot 2 \cdot 3 \cdot 4} = 35$$

Example 1.3.1. *Assuming that every offspring has a 50–50 chance of being male (or female, for that matter), what is the probability that exactly three of a family's five children are male?*

Here $n = 5$, $k = 3$, and $p = 0.5$. Accordingly, the probability in question is

$$b(3; 5, 0.5) = \binom{5}{3} \cdot 0.5^3 \cdot 0.5^2 = 10 \cdot 0.125 \cdot 0.25 = 0.3125 = 31.25\%$$

Example 1.3.2. *What is the probability of getting exactly three 6s in seven throws of a single die?*

Strictly speaking, the toss of a die is not a binomial trial because it has six possible outcomes and not the required two. However, this can be fixed by considering the outcome of a 6 as a success and the outcome of *not a 6* as a failure. This converts the toss of a die into a binomial trial with $p = \dfrac{1}{6}$ and, of course, $1 - p = 1 - \dfrac{1}{6} = \dfrac{5}{6}$. Here $n = 7$, $k = 3$, $n - k = 4$, and consequently the required probability is

$$b(3; 7, \tfrac{1}{6}) = \binom{7}{3} \left(\frac{1}{6} \right)^3 \left(\frac{5}{6} \right)^4 = \frac{7 \cdot 6 \cdot 5}{1 \cdot 2 \cdot 3} \left(\frac{1}{6} \right)^3 \left(\frac{5}{6} \right)^4 = 0.0781 \approx 7.8\%$$

Example 1.3.3. *Suppose 30% of the population of a certain city are fans of that city's pro basketball team. Suppose further that an advertising company's researcher addresses 10 people at random while walking around the city and asks them whether they are indeed fans. What is the probability that exactly three of the respondents turn out to be fans?*

The choice of a person to interview is similar to the throw of a die. In each case we are looking at independent repetitions of a random event with two possible outcomes of which one is considered to be a success: finding a fan in one case and throwing a 6 in the other. Reasoning by analogy, we find that the probability of encountering exactly three fans is

$$b(3; 10, 0.3) = \binom{10}{3} \cdot 0.3^3 \cdot (1 - 0.3)^7 = \frac{10 \cdot 9 \cdot 8}{1 \cdot 2 \cdot 3} \cdot 0.3^3 \cdot 0.7^7 = 0.2668 \approx 26.7\%$$

Similarly, if 45% of the population of this city is male and a surveyor phones eight randomly selected numbers, then the probability of three people who answer the call being male is

$$b(3; 8, 0.45) = \binom{8}{3} \cdot 0.45^3 \cdot (1 - 0.45)^{8-3} = \frac{8 \cdot 7 \cdot 6}{1 \cdot 2 \cdot 3} \cdot 0.45^3 \cdot 0.55^5 = 0.2568 \approx 25.7\%$$

These last examples, although seemingly practical, are in fact unrealistic because we are very unlikely to ever possess enough information to allow us to determine the precise proportion of fans or males in a town. In practice, one is more often interested in the reverse question of what information the results of the sampling provide regarding the proportion of fans in the entire population. This question is dealt with in Section 2.5. At this point, however, it is possible to garner some information regarding the related issue of the likelihood of a sample being in some sense misleading.

Example 1.3.4. *One would expect a random sample of 10 inhabitants of this town to contain 3 fans, and hence such a sample might give the surveyor the wrong impression if it contained no more than 1 fan. What is the probability of this happening?*

Because "no more than 1 fan" is tantamount to 0 or 1 fan, the probability of this happening is

$$b(0;10,0.3) + b(1;10,0.3) = \binom{10}{0} \cdot 0.3^0 \cdot 0.7^{10} + \binom{10}{1} \cdot 0.3^1 \cdot 0.7^9$$
$$\approx 0.0282 + 0.1211 = 0.1493 \approx 14.9\%$$

Example 1.3.5. *A sample of 10 inhabitants could be considered to be misleading if it contains too many, say 5 or more, fans. The probability of this happening is*

$$b(5;10, 0.3) + b(6; 10, 0.3) + b(7; 10, 0.3) + b(8; 10, 0.3)$$
$$+ b(9; 10, 0.3) + b(10; 10, 0.3)$$

$$= \binom{10}{5} \cdot 0.3^5 \cdot 0.7^5 + \binom{10}{6} \cdot 0.3^6 \cdot 0.7^4 + \binom{10}{7} \cdot 0.3^7 \cdot 0.7^3 + \binom{10}{8} \cdot 0.3^8 \cdot 0.7^2$$

$$+ \binom{10}{9} \cdot 0.3^9 \cdot 0.7^1 + \binom{10}{10} \cdot 0.3^{10} \cdot 0.7^0$$

$$\approx 0.1029 + 0.0367 + 0.0090 + 0.0038 + 0.0001 + 0.0000 = 0.150 = 15\%$$

Put together, the above examples say that $14.9\% + 15.3\% \approx 30\%$ of the random samples could mislead the unwary surveyor, an uncomfortably high proportion.

Such probabilities are useful and need to be computed frequently. The above described method is unfortunately very unwieldy. For instance, suppose it is necessary to compute the probability of a sample of 20 inhabitants containing at least 10 fans (as opposed to the expected 6). The answer can be written as the sum

$$b(10; 20, 0.3) + b(11; 20, 0.3) + \cdots + b(19; 20, 0.3) + b(20; 20, 0.3)$$

Depending on the calculator at one's disposal as well as on one's skills, the evaluation of this sum could vary between a triviality and a nuisance. Before the advent of the electronic calculators, however, such problems constituted unpleasant and lengthy chores. They did arise often enough, however, to motivate the creation of appropriate tables. The use of these tables requires a new symbol. If a binomial process involves n repetitions of an elementary trial with a probability of success p, then the probability of at most k successes is denoted by

$$B(k; n, p) = b(0; n, p) + b(1; n, p) + \cdots + b(k; n, p) \qquad (1.1)$$

Some of these values are tabulated in Table B. Accordingly, the probability that the sample of 20 inhabitants encounters at most 4 fans is

$$B(4; 20, 0.3) = 0.2375 = 23.75\%$$

Similarly, the probability that a sampling of 15 inhabitants yields at most 7 fans is $B(7; 15, 0.3) = 0.9500 \approx 95\%$. Table B can be used to derive other probabilities as well.

Example 1.3.6. *Suppose a random sample of 15 of the inhabitants of the town of the above examples are interviewed. Determine the probabilities of the following events:*

a. *At least seven are fans.* The complementary event is that at most six are fans and has probability $B(6; 15, 0.3)$. The required probability therefore is

$$1 - B(6; 15, 0.3) = 1 - 0.8689 = 0.1311 \approx 13.1\%$$

b. *The number of fans is between four and seven (inclusive).* This probability is

$$B(7; 15, 0.3) - B(3; 15, 0.3) = 0.9500 - 0.2969 = 0.6513 \approx 65.3\%$$

c. *The number of fans is exactly five.* This probability is

$$B(5; 15, 0\ 3) - B(4; 15, 0.3) = 0.7216 - 0.5155 = 0.2061 \approx 20.6\%$$

Example 1.3.7. *In any match between the baseball teams NL and AL, the first has the probability 0.6 of winning. Suppose the two teams meet in a World Series wherein a team must win four games to emerge victorious. What is the probability of team NL winning the series?*

The baseball World Series is a binomial process with $n = 7$, $k \geq 4$, and $p = 0.6$. Hence, the desired probability is

$$1 - B(3; 7, 0.6) = 1 - 0.2898 = 0.7102 \approx 71\%.$$

Example 1.3.8. *In a certain baseball World Series, the two teams are considered to be equally strong. Suppose the NL team wins the first two games of the series. What is the probability that this team will go on to win the entire series?*

To win the series, the team NL needs to win at least two of the remaining five games. It is therefore facing a binomial process with $n = 5$, $k \geq 2$, and $p = 0.5$ and the probability of its winning is

$$1 - B(1; 5, 0.5) = 1 - 0.1875 = 0.8125 = 81.25\%$$

EXERCISES 1.3

Use the formula for $b(k; n, p)$ to compute the probabilities required in Exercises 1–7.

1. A fair die is tossed five times. What is the probability of obtaining

    a) no 6s

    b) exactly one 6

    c) exactly two 6s

    d) exactly five 6s

2. A fair die is tossed six times. What is the probability of obtaining

    a) no 5s

    b) exactly one 5

    c) exactly two 5s

    d) exactly six 5s

3. The probability that a patient recovers from a certain disease is 0.75. If a hospital has seven patients with this disease, what is the probability of the following events:

    a) none recover

    b) exactly one recovers

    c) exactly five recover

    d) all recover

4. In a certain city the need for money to buy drugs is stated as the reason for two thirds of all burglaries. Find the probability that among the next eight burglaries the number of those due to this need is

    a) 0

    b) exactly six

    c) exactly eight

5. It is known that one third of trucks will experience a tire blowout in a certain strenuous test. Suppose five trucks are tested. Find the probability that the number of trucks that experience a blow out is

    a) 0

    b) exactly two

    c) exactly five

6. It is estimated that a certain flu vaccine is 85% effective. If 13 people are inoculated, find the probability that the number of successful vaccinations among them is

   a) exactly 1

   b) exactly 13

7. Of the cars passing through a toll booth, 15% are from outside the state. Find the probability that of the next six cars the number of out-of-staters is

   a) 0

   b) exactly one

   c) exactly two

Use Formula (1.1) and Table B to compute the probabilities required in Exercises 8–14.

8. A fair coin is tossed 15 times. What is the probability of obtaining

   a) at most 9 heads

   b) at least 9 heads

   c) exactly 9 heads

   d) between 8 and 12 heads inclusive

9. A fair coin is tossed 20 times. What is the probability of obtaining

   a) at most 12 heads

   b) at least 12 heads

   c) exactly 12 heads

   d) between 8 and 12 heads inclusive

10. The probability that a patient recovers from a certain disease is 0.7. If a hospital has 12 patients with this disease, what is the probability of the following events:

    a) at most 5 recover

    b) at least 5 recover

    c) exactly 5 recover

    d) between 5 and 10 (inclusive) recover

11. In a certain city the need for money to buy drugs is stated as the reason for 80% of all burglaries. Find the probability that among the next eight burglaries the number of those due to this need is

   a)   at most six

   b)   at least six

   c)   exactly six

   d)   between four and six inclusive

12.   It is known that 30% of trucks will experience a tire blowout in a certain strenuous test. Suppose five trucks are tested. Find the probability that the trucks that fail is

   a)   at most two

   b)   at least two

   c)   exactly two

   d)   between one and three inclusive

13.   It is estimated that a certain flu vaccine is 90% effective. If 12 people are inoculated, find the probability that the number of successful vaccinations among them is

   a)   at most 10

   b)   at least 10

   c)   exactly 10

   d)   between 8 and 11 inclusive

14.   Of the cars passing through a toll booth, 20% are from outside the state. Find the probability that of the next 16 cars the number of out-of-staters is

   a)   at most 6

   b)   at least 6

   c)   exactly 6

   d)   between 4 and 8 inclusive

1.4   The Normal Approximation

Realistic polls may lead to such numbers as the probability that a random sample of 100 inhabitants contains at most 35 fans, which, according to the previous section, is $B(35; 100, 0.4)$. There is no table for this at the end because the listing of the entries for n up to 100 is impractical, as is indicated by the rate at which the numbers of displayed entries for $n = 1, 2, ..., 20$ increase. On the other hand, Formula (1.1) yields the lengthy sum

$$\binom{100}{0} \cdot 0.4^0 \cdot 0.6^{100} + \binom{100}{1} \cdot 0.4^1 \cdot 0.6^{99} + \cdots + \binom{100}{34} \cdot 0.4^{34} \cdot 0.6^{66} + \binom{100}{35} \cdot 0.4^{35} \cdot 0.6^{65}$$

which is not easily calculated either by hand or with a hand-held calculator. Of course, there are mathematical computer applications that will compute this sum in a jiffy, but even these will run into trouble with sums of binomial probabilities that arise in more technical circumstances. The mathematicians of the 18th century, who had no calculators available, resorted instead to their brains and in the process of overcoming this particular computational difficulty discovered a phenomenon that eventually became the cornerstone of the whole theory of statistics. Although this discovery was first phrased algebraically, it is more easily explained geometrically in terms of graphs.

Figures 1.4 to 1.7 display the probabilities associated with the flipping of a fair coin. These are binomial experiments with $p = 0.5$ and $n = 10, 20, 50$, and 100, respectively. In these bar graphs the probability of each event of k successes is represented by the height of a rectangular strip of width 1 centered over k. Not surprisingly, all these

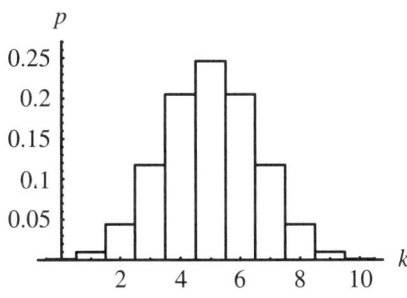

Figure 1.4 $n = 10, p = 0.5$

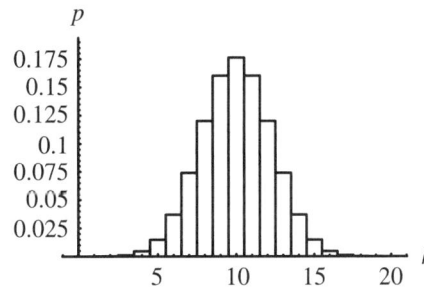

Figure 1.5 $n = 20, p = 0.5$

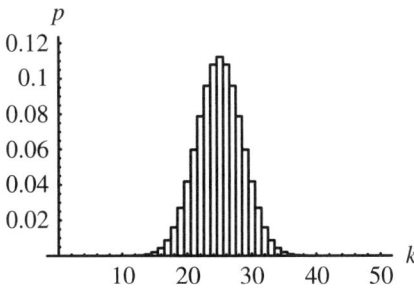

Figure 1.6 $n = 50, p = 0.5$

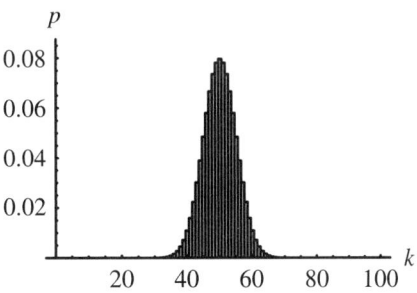

Figure 1.7 $n = 100, p = 0.5$

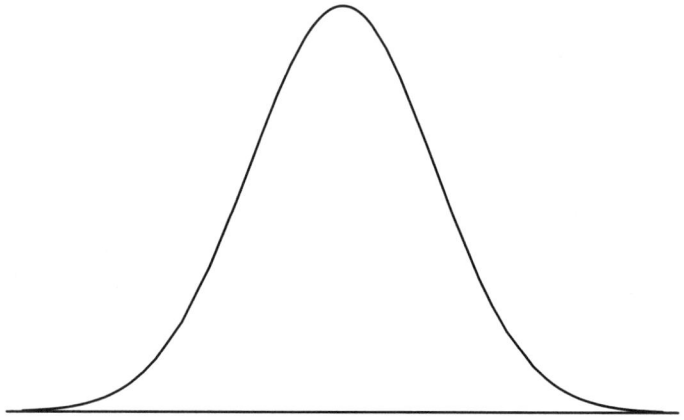

Figure 1.8   *The bell-shaped curve.*

graphs have the same general shape, which is popularly known as the bell-shaped curve and is portrayed in Figure 1.8.

Figures 1.9 to 1.12 demonstrate how the bell-shaped curve can be fitted to the binomial probabilities of Figures 1.4 to 1.7 previously. Figures 1.13 to 1.20 demonstrate

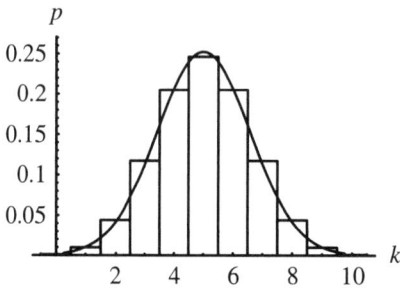

Figure 1.9 $n = 10, p = 0.5$

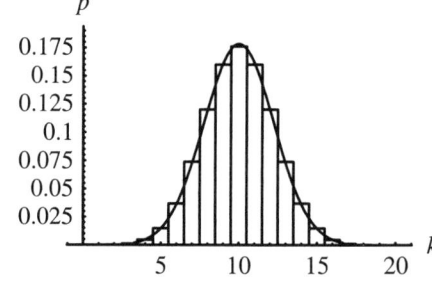

Figure 1.10 $n = 20, p = 0.5$

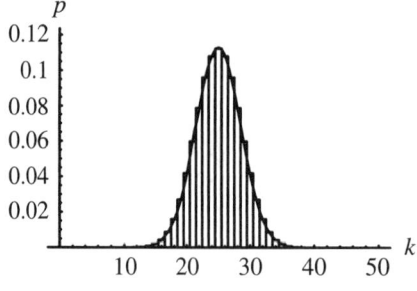

Figure 1.11 $n = 50, p = 0.5$

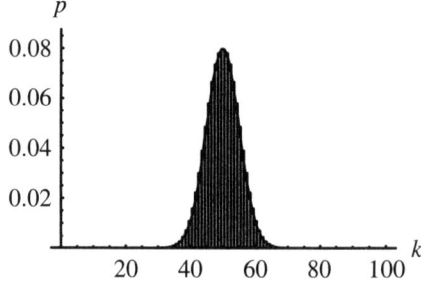

Figure 1.12 $n = 100, p = 0.5$

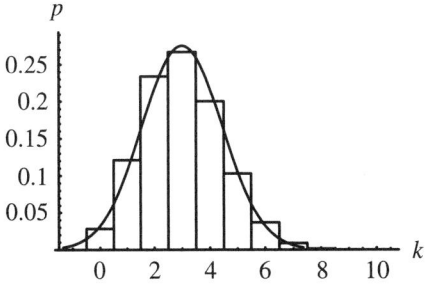

Figure 1.13 $n = 10,\ p = 0.3$

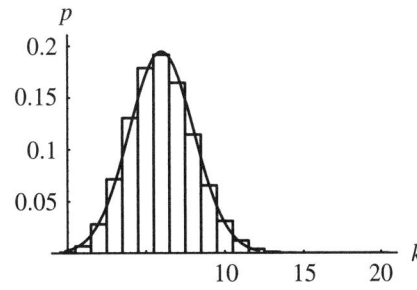

Figure 1.14 $n = 20,\ p = 0.3$

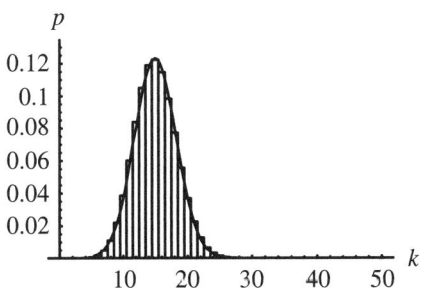

Figure 1.15 $n = 50,\ p = 0.3$

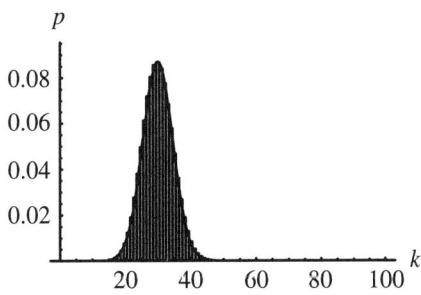

Figure 1.16 $n = 100,\ p = 0.3$

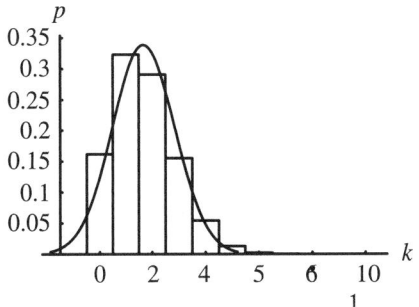

Figure 1.17 $n = 10,\ p = \dfrac{1}{6}$

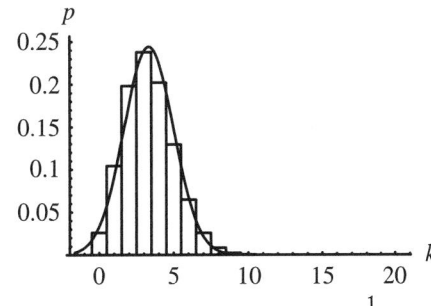

Figure 1.18 $n = 20,\ p = \dfrac{1}{6}$

Figure 1.19 $n = 50$, $p = \dfrac{1}{6}$

Figure 1.20 $n = 100$, $p = \dfrac{1}{6}$

that the symmetric bell-shaped curve can also be fitted to the asymmetric binomial probabilities associated with $p = 0.3$ and $p = \dfrac{1}{6}$. It is clear that for reasonably large n, that is, when the experiment is repeated many times, the asymmetries are so small as to be negligible.

The approximating bell-shaped curves in Figures 1.8 through 1.20 are all different and are all called *normal curves*. They differ in their widths and both the location and height of their peaks. Nevertheless, in 1735 Abraham de Moivre (1667–1754) demonstrated that all these curves were identical in the sense that any one could be transformed into any other by a rescaling of the axes and/or a horizontal shift. Moreover, this equivalence greatly facilitated the computation of binomial probabilities.

Consider, for the purpose of an illustration, the problem of computing B(4; 10, 0.4) (for which Table B yields the value of 0.633). Because each of the strips in the graphic representation of the binomial probabilities has width 1, the probability represented by any strip actually equals its area. Hence, the probability B(4; 10, 0.4) is represented by the shaded region in Figure 1.21.

This area is approximately equal to the shaded region in Figure 1.22. Because all normal curves are essentially the same, it suffices to tabulate such areas for a single

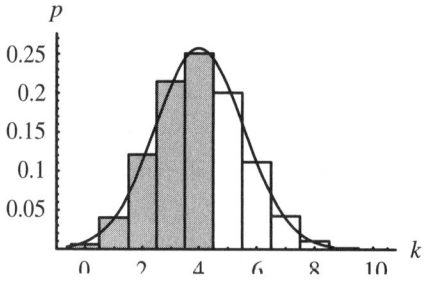

Figure 1.21   A binomial probability

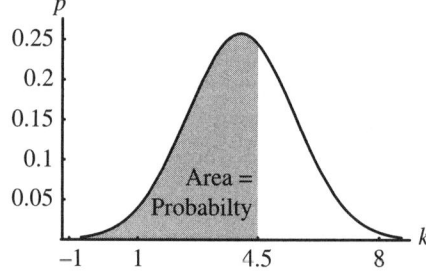

Figure 1.22   An approximation to a probability

standard normal curve and then provide a method for transporting this information to arbitrary normal curves. The *standard* normal curve is the graph of the function

$$y = \frac{e^{-x^2/2}}{\sqrt{2\pi}}$$

which is displayed in Figure 1.23. Here π is the well-known ratio of the circumference of a circle to its diameter, with value 3.14159 ..., whereas e is another important mathematical constant with value 2.71828 ... . The area of the region underneath the standard normal curve and to the left of the number x is denoted by $A(x)$, (Figure 1.24) and its values are tabulated in Table A. Table A shows that $A(1) = 0.8413$,

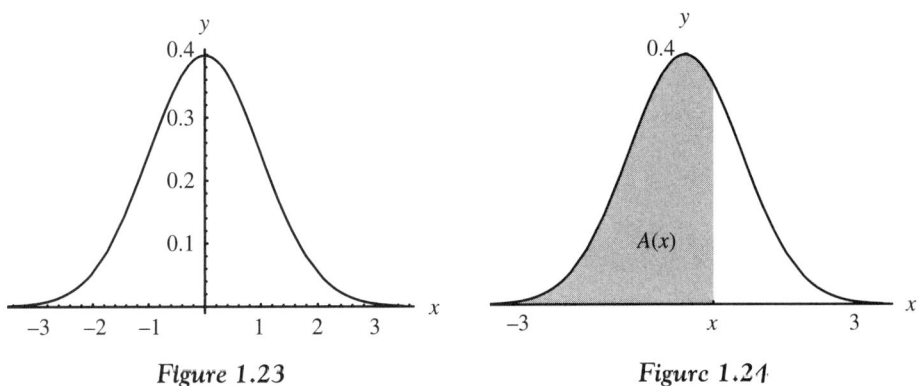

Figure 1.23   Figure 1.24

$A(0.5) = 0.6915$, and $A(-1.17) = 0.1210$. The evaluation of the cumulative binomial probabilities is given by the formula

$$B(k; n, p) = A\left(\frac{k + 0.5 - pn}{\sqrt{p(1-p)n}} \right)$$

Accordingly,

$$B(4; 20, 0.3) = A\left(\frac{4.5 - 0.3 \cdot 20}{\sqrt{0.3 \cdot 0.7 \cdot 20}} \right) = A(-0.73) = 0.2327 \approx 23.3\%$$

and

$$B(7; 15, 0.3) = A\left(\frac{7.5 - 0.3 \cdot 15}{\sqrt{0.3 \cdot 0.7 \cdot 15}} \right) = A(1.69) = 0.9545 \approx 95.5\%$$

both of which agree closely with the previously obtained results (just above Example 1.3.6) of 23.75% and 95%, respectively. Moreover, the evaluation of untabulated binomial probabilities is now also feasible. For example,

$$B(35;100,0.4) = A\left(\frac{35.5 - 0.4 \cdot 100}{\sqrt{0.4 \cdot 0.6 \cdot 100}}\right) = A(-0.92) = 0.1788 \approx 17.9\%$$

Example 1.4.1. *It is known that the recovery rate for a certain rare disease is seven out of eight. Suppose 25 people contract this disease. What is the probability that (a) no more than 20 recover, (b) at least 20 recover, (c) from 20 to 23 recover, (d) exactly 21 recover?*

(a) This is $B(20;25,7/8) = A\left(\dfrac{20.5 - (7/8) \cdot 25}{\sqrt{(7/8) \cdot (1/8) \cdot 25}}\right) = A(-0.83) = 0.2033 \approx 20.3\%$

(b) This is $1 - B(19;25,7/8) = 1 - A\left(\dfrac{19.5 - (7/8) \cdot 25}{\sqrt{(7/8) \cdot (1/8) \cdot 25}}\right) = 1 - A(-1.44)$

$$= 1 - 0.0749 = 0.9251 \approx 92.5\%$$

(c) This is $B(23;25,7/8) - B(19;25,7/8) =$

$$A\left(\frac{23.5 - (7/8) \cdot 25}{\sqrt{(7/8) \cdot (1/8) \cdot 25}}\right) - A\left(\frac{19.5 - (7/8) \cdot 25}{\sqrt{(7/8) \cdot (1/8) \cdot 25}}\right) = A(0.98) - A(-1.44)$$

$$= 0.8365 - 0.0749 = 0.7616 \approx 76.2\%$$

(d) This is $B(21;25,7/8) - B(20;25,7/8)$

$$= A\left(\frac{21.5 - (7/8) \cdot 25}{\sqrt{(7/8) \cdot (1/8) \cdot 25}}\right) - A\left(\frac{20.0 - (7/8) \cdot 25}{\sqrt{(7/8) \cdot (1/8) \cdot 25}}\right) = A(-0.23) - A(-0.83)$$

$$= 0.4090 - 0.2033 = 0.2057 \approx 20.6\%$$

In the second half of the 19th century, the Englishman Sir Francis Galton did much to popularize the normal curve among the scientists of the day. As part of this effort, in 1873 he constructed a device, called the *quincunx*, that strikingly demonstrated the ability of this curve to approximate binomial probabilities. In this device, shot was poured through an array of pins and because each time a shot hit a pin it would go to the right or to the left with an equal probability of 0.5, the paths taken by these shots simulated a binomial experiment (success = right, say, and failure = left). The shots then accumulated in compartments at the bottom of the quincunx where their outline would resemble a normal curve. Quincunxes are displayed in many science museums, and an electronic simulation can be found at the RAND corporation's website at *www.rand.org/centers/stat/applets/clt.html*, among others.

EXERCISES 1.4

1. A coin is tossed 200 times. What is the probability of obtaining
    a) at most 90 heads
    b) at least 90 heads
    c) exactly 100 heads
    d) between 90 and 110 heads (inclusive)

2. A die is tossed 300 times. What is the probability of obtaining
    a) at most 40 ones
    b) at least 40 ones
    c) exactly 50 ones
    d) between 40 and 60 (inclusive) ones

3. It was demonstrated in Section 1.1 that if three dice are rolled, then the probability of the faces adding up to 9 is $\frac{25}{216}$. Suppose three such dice are rolled 200 times. Find the probability of obtaining this sum
    a) at most 25 times
    b) at least 25 times
    c) between 20 and 30 times (inclusive)
    d) exactly 25 times

4. It was demonstrated in Section 1.1 that if three dice are rolled, then the probability of the faces adding up to 10 is $\frac{27}{216}$. Suppose three such dice are rolled 200 times. Find the probability of obtaining this sum
    a) at most 27 times
    b) at least 27 times
    c) between 20 and 30 times (inclusive)
    d) exactly 27 times

5. Of the cars passing through a toll booth, 25% are from outside the state. Find the probability that of the next 1000 cars the number of out-of-staters is
    a) at most 260
    b) at least 260
    c) exactly 260
    d) between 240 and 280 (inclusive)

6. A fair coin is tossed 150 times. What is the probability of obtaining

    a) at most 60 heads

    b) at least 60 heads

    c) exactly 75 heads

    d) between 60 and 80 heads (inclusive)

7. A fair coin is tossed 30 times. What is the probability of obtaining

    a) at most 12 heads

    b) at least 12 heads

    c) exactly 15 heads

    d) between 8 and 12 heads (inclusive)

8. The probability that a patient recovers from a certain disease is 0.7. If a hospital has 25 patients with this disease, what is the probability of the following events:

    a) at most 12 recover

    b) at least 12 recover

    c) exactly 15 recover

    d) between 12 and 18 recover (inclusive)

9. In a certain city the need for money to buy drugs is stated as the reason for 80% of all burglaries. Find the probability that among the next 200 burglaries the number of those due to this need is

    a) at most 150

    b) at least 150

    c) exactly 160

    d) between 150 and 170 (inclusive)

10. It is known that 25% of trucks will experience a tire blowout in a certain strenuous test. Suppose 35 trucks are tested. Find the probability that the number of trucks that experience a blowout is

    a) at most 12

    b) at least 12

    c) exactly 10

    d) between 10 and 12 (inclusive)

11. It is estimated that a certain flu vaccine is 90% effective. If 1300 people are inoculated, find the probability that the number of successful vaccinations among them is

   a) at most 1200

   b) at least 1200

   c) exactly 1200

   d) between 1150 and 1250 (inclusive)

1.5   Activities/Websites

1. Activate the quincunx at *www.rand.org/centers/stat/applets/clt.html*. Let it run for a few minutes until the region below the normal curve is almost covered and then print the page. Does the outcome support the contention that the normal curve can be used to approximate binomial probabilities?

2. A contestant on a TV show faces three doors: One hides a car and the other two hide goats. The contestant selects a door, whereupon the host, who knows exactly which door hides what prize, opens one of the other two doors to reveal a goat. The contestant is then given the choice of staying with his or her original selection or switching to the third door. Should the contestant switch or not?

   a) Go to the website *http://www.shodor.org/interactivate/activities/monty3/* and obtain the answer by running some electronic simulations.

   b) Find a theoretical answer (on the web, if need be).

   c) Do the simulations and the theoretical answer agree?

3. Theoretical considerations indicate that in a group of seven people, the probability of some two of them being born in the same day of the month is slightly greater than 50%. Check this out by dividing your classmates into groups of seven and counting the groups with multiple birthdays. Use the phrase "birthday problem" to find some relevant information on the web.

Chapter 2
Statistics

All measurements, scientific or otherwise, are to be taken with a grain of salt because they involve some errors. Although these inevitable errors were first regarded as merely a nuisance, they eventually became the object of scientific inquiry themselves. It was discovered that measurement errors follow probabilistic rules that are similar to those that govern binomial processes. This eventually led to myriads of applications of the normal curve in both the sciences and everyday life.

2.1 Summarizing Data

Modern science and technology are data driven, and all theories, no matter how elegant or "common sensical," must be backed up by facts. Unfortunately, raw data often come in bewildering quantities, and in such cases it is easy to miss significant patterns because of the profusion of details. For this reason it is customary to summarize such data by means of quantitative parameters, groupings, graphs, or any combination thereof.

The quantitative parameters used to describe data are of two kinds: *measures of central tendencies* and *measures of variability*. Those of the former kind attempt to pinpoint the "center" of the data and those of the latter pertain to their dispersion.

2.1.1 Central Tendencies

The most commonly used measure of central tendency is the *average*, also known as the *mean* or the *arithmetic mean*. The average of the data $x_1, x_2, ..., x_n$ is denoted by $\bar{x}$ and equals

$$\bar{x} = \frac{x_1 + x_2 + \cdots + x_n}{n}$$

Because the computation of the average calls for a division, it generally requires more digits behind the decimal point than the data, and we agree to stop after one additional digit has been obtained. Thus, the mean of the data 3, 7, 6 is 5.33... ≈ 5.3, that of the data 3, 7, 7 is 5.66 ... ≈ 5.7, and that of 1.3, 1.7, 1.7 is 1.566 ... ≈ 1.57.

The *median* of a set of data is almost as commonly computed as its mean. This measure is denoted by $\tilde{x}$, and its determination calls for first ordering the data in increasing order so that it may be assumed that $x_1 \leq x_2 \leq \cdots \leq x_n$ and then setting

$$\tilde{x} = \begin{cases} x_{(n+1)/2} & \text{if } n \text{ is odd} \\ \dfrac{x_{n/2} + x_{(n/2)+1}}{2} & \text{if } n \text{ is even} \end{cases}$$

For example, if the data are 15.7, 4.9, 3.2, 2.7, 4.1, 10.5, 8.9, then they need to be rearranged as 2.7, 3.2, 4.1, 4.9, 8.9, 10.5, 15.7. Because there are seven data, the median is

$$\tilde{x} = x_{(7+1)/2} = x_{8/2} = x_4 = 4.9$$

On the other hand, if there are eight data—15.7, 4.9, 3.2, 13.4, 2.7, 4.1, 10.5, 8.9— then they are rearranged as 2.7, 3.2, 4.1, 4.9, 8.9, 10.5, 13.4, 15.7 and

$$\tilde{x} = \frac{x_{8/2} + x_{(8/2)+1}}{2} = \frac{x_4 + x_5}{2} = \frac{4.9 + 8.9}{2} = 6.9$$

Loosely speaking, the median is a number that splits the data by size into two groups of equal magnitudes. Unfortunately, all too often this is imprecise. The data 1, 2, 2, 3, 3, 3, 3, 5, 9 whose median is 3 has 7/9 of its members at 3 or below and 8/9 of its members at 3 or above, both of which fractions are considerably bigger than 1/2. More precisely, it can be asserted that for *at least* half of the data

$$x_i \leq \tilde{x}$$

and for *at least* half of the data

$$x_i \geq \tilde{x}$$

The *mode* of a set of data is its most frequently occurring member. Thus, the mode of the data 1, 2, 2, 3, 3, 3, 3, 5, 9 is 3. Whereas every set of data has both a mean and a median, its mode need not be well defined. Thus, the data 1, 2, 2, 2, 3, 3, 4, 5, 5, 5 has no mode (it is sometimes said to be *bimodal*).

The mode is considered to be the least *sensitive* (or, equivalently, most *robust*) of these three measures of centrality because there are many ways to modify a data set without affecting its mode. For example, the data set

$$1, 2, 2, 2$$

and its extension

$$1, 2, 2, 2, 3, 3, 4, 4, 5, 5, 6, 6, 7, 7$$

clearly have the same mode. The median is also quite insensitive. For example, any modification of all the data (except the median itself) that does not affect their relative magnitudes will not affect the median. Such, for example, is the case when the data

$$1, 2, 3, 4, 5, 6, 7$$

is modified to

$$0.001, 0.02, 0.3, 4, 50, 600, 7000$$

The mean is the most sensitive of these measures. Any change in the value of some of the data is very likely to affect the value of their average.

Sensitivity and its opposite, robustness, have both advantages and disadvantages. Those who wish to protect against outliers—data that lie so far from the rest that one has good reason for discounting their validity or relevance—are advised to use the median because its value is mostly independent of such accidents. If, however, all the data are judged to be equally trustworthy, then the average is considered to be the better representative. The mode should be used when it is crucial that the representative number should be an integer. Thus, the typical family in a certain town has two or three children, but never 2.6 children. Similarly, biologists, when describing the typical nest of a species of birds, use the mode for the number of eggs.

2.1.2   Variability

The second most important aspect of a set of data is its *variability*, or *spread*. One easy way of measuring this is simply to look for the difference between the largest and smallest data; this is called the *range* of the data. For example, the range of the data 4.7, 2.3, 1.6, 8.3, 5.2, 7.4, 2.5 is

$$8.3 - 1.6 = 6.7$$

The range is not very sensitive and, for example, does not distinguish between the two data sets

$$1, 2, 3, 4, 5, 6, 7, 8, 9 \quad \text{and} \quad 1, 5, 5, 5, 5, 5, 5, 5, 9$$

the first of which clearly has much more variability than the second. Sophisticated use of data, as is illustrated below, calls for a more intricate measure. Our strategy is to define the variability of the data in terms of their distances from their common mean, i.e., in terms of $x_i - \bar{x}$. However, because the sum

$$(x_1 - \bar{x}) + (x_2 - \bar{x}) + \cdots + (x_n - \bar{x}) \tag{2.1}$$

is always zero (see Exercises 2.1, numbers 15 and 16), it is necessary to make these terms positive so that they do not cancel each other out. For reasons that are explained later, it has become customary to square the individual differences and work with

$$(x_1 - \overline{x})^2 + (x_2 - \overline{x})^2 + \cdots + (x_n - \overline{x})^2$$

This sum is then averaged to

$$\frac{(x_1 - \overline{x})^2 + (x_2 - \overline{x})^2 + \cdots + (x_n - \overline{x})^2}{n}$$

because it is necessary to make sure that data sets such as

$$1, 3 \quad \text{and} \quad 1, 1, 1, 1, 3, 3, 3, 3$$

have equal variability (see Exercises 2.1, numbers 1 and 2). Finally, to offset the effect of the squaring, a square root is taken and the *standard deviation* of the data x_1, x_2, ..., x_n is defined as

$$\sqrt{\frac{(x_1 - \overline{x})^2 + (x_2 - \overline{x})^2 + \cdots + (x_n - \overline{x})^2}{n}}$$

For example, to compute the standard deviation of the data 5, 7, note that their mean $\overline{x}$ is 6, so their standard deviation is

$$\sqrt{\frac{(5-6)^2 + (7-6)^2}{2}} = \sqrt{\frac{(-1)^2 + 1^2}{2}} = \sqrt{\frac{2}{2}} = \sqrt{1} = 1$$

It is common to denote the standard deviation by the letter s. Thus, for the data 2, 3, 5, 10, we have

$$\overline{x} = \frac{2 + 3 + 5 + 10}{4} = \frac{20}{4} = 5$$

and

$$s^2 = \frac{(2-5)^2 + (3-5)^2 + (5-5)^2 + (10-5)^2}{4} = \frac{9 + 4 + 0 + 25}{4} = \frac{38}{4}$$

so that

$$s = \sqrt{\frac{38}{4}} \approx 3.1$$

Similarly, for the data 4.7, 2.3, 1.6, 8.3, 5.2, 7.4, 2.5 we have

$$\overline{x} = \frac{4.7 + 2.3 + 1.6 + 8.3 + 5.2 + 7.4 + 2.5}{7} = \frac{32}{7} \approx 4.57$$

and

$$s^2 = \frac{1}{7}[(4.7 - 4.57)^2 + (2.3 - 4.57)^2 + (1.6 - 4.57)^2 + (8.3 - 4.57)^2 +$$

$$(5.2 - 4.57)^2 + (7.4 - 4.57)^2 + (2.5 - 4.57)^2] \approx 5.7992$$

so that

$$s = \sqrt{5.7992} \approx 2.41$$

2.1.3   Groupings and Histograms

Another way of coping with the obfuscation caused by the large size of a data set is to subdivide it into convenient groups and then represent those by means of bar graphs. The process of grouping has not been standardized and generally depends on the data as well as subjective considerations. Two procedures are described here.

Consider the following data that might describe the ages of the students in a certain university class:

$$\{21, 20, 21, 22, 21, 18, 20, 19, 19, 21, 19, 21, 20, 21, 20, 20, 23, 20, 20, 22,$$
$$20, 21, 21, 19, 22, 22, 17, 21, 20, 21, 21, 23, 19, 22, 21, 18, 19, 19, 21\}.$$

The *frequency table* of these data is obtained by recording the frequency f_i with which the ith age occurs rather than listing them all (Table 2.1). It is customary to also include in such tables the *relative frequency*:

$$r_i = \frac{f_i}{n}$$

where n denotes the total number of data (Table 2.1). The *histogram* of a set of data is a visual presentation of its frequencies as a bar graph (Fig. 2.1). The average and the standard deviation of a data set are easily computed from its frequency table:

$$\bar{x} = r_1 x_1 + r_2 x_2 + \cdots r_k x_k$$

$$s = \sqrt{r_1 (x_1 - \bar{x})^2 + r_2 (x_2 - \bar{x})^2 + \cdots + r_k (x_k - \bar{x})^2} \tag{2.2}$$

Thus, for the data of Table 2.1:

$$\bar{x} = 0.03 \cdot 17 + 0.05 \cdot 18 + 0.18 \cdot 19 + 0.23 \cdot 20 + 0.33 \cdot 21 + 0.13 \cdot 22 + 0.05 \cdot 23 \approx 20.4$$

$$s^2 = [0.02(17 - 20.4)^2 + 0.05(18 - 20.4)^2 + 0.18(19 - 20.4)^2 + 0.23(20 - 20.4)^2$$
$$+ 0.33(21 - 20.4)^2 + 0.13(22 - 20.4)^2 + 0.05(23 - 20.4)^2] \approx 1.814$$

so that

$$s \approx \sqrt{1.814} \approx 1.35$$

Group index i	Age	Frequency f_i	Relative frequency r_i
1	17	1	0.03
2	18	2	0.05
3	19	7	0.18
4	20	9	0.23
5	21	13	0.33
6	22	5	0.13
7	23	2	0.05
		39	1

Table 2.1   Ages in a university class

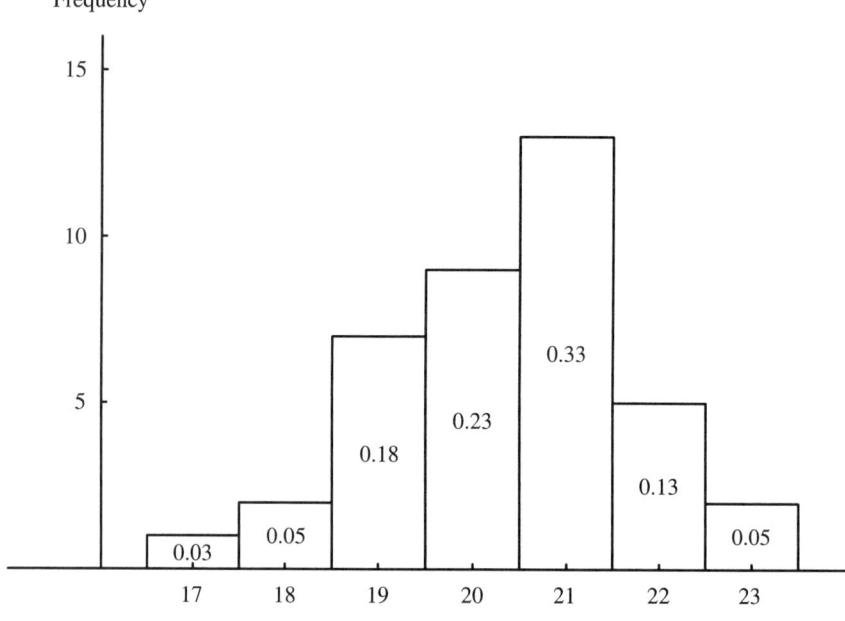

Figure 2.1   A histogram for the ages in a university class.

When the data are spread out more finely, each number is likely to occur only once and so it is better to group them into classes. Consider, for example, the following data that might represent the ages of the students attending an adult education class, where the students' ages are likely to vary much more than in a university class (Table 2.2):

{26, 49, 43, 42, 46, 60, 34, 43, 41, 25, 20, 54, 46, 41, 56, 76, 33, 47, 56, 46, 34, 38, 45, 25, 33, 36, 38, 40, 65, 35, 49, 41, 25, 37, 43, 53, 51, 32, 18, 34, 39, 60, 19, 48, 46}

Group index i	Age groups	Frequency f_i	Relative frequency r_i
1	10 – 19	2	0.04
2	20 – 29	5	0.11
3	30 – 39	12	0.27
4	40 – 49	17	0.38
5	50 – 59	5	0.11
6	60 – 69	3	0.07
7	70 – 79	1	0.02
		45	1

Table 2.2   Ages in an adult education class

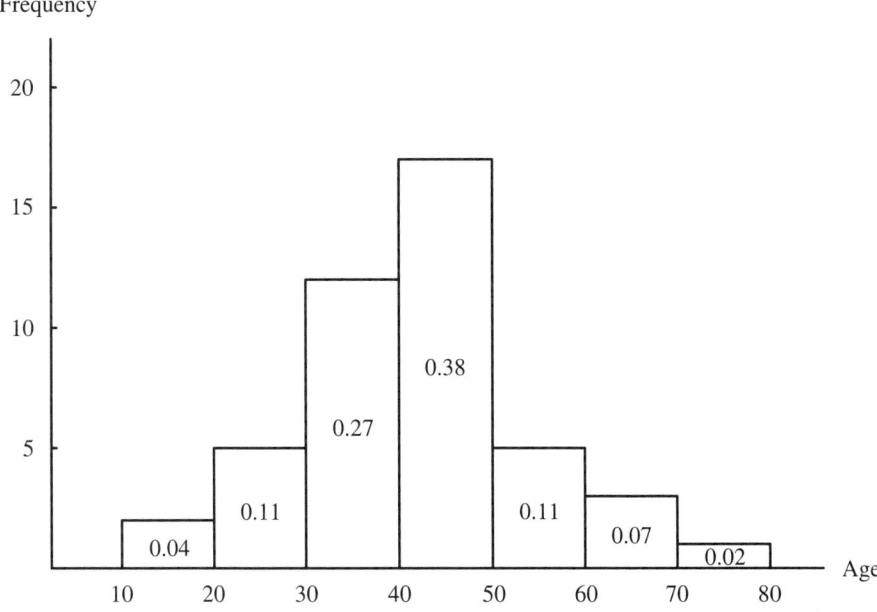

Figure 2.2   *A histogram for the ages in an adult education class.*

Given that these numbers represent ages and that people tend to think of age in terms of decades, it is reasonable to group them as in Table 2.2.

Approximations to the mean and standard deviation of a data set are easily derived from Figure 2.2 by means of formulas analogous to those of (2.2) earlier. The average of the limits of the ith data group is called its *mark* and is denoted by m_i. These marks replace the actual data in (2.2) so that we obtain

$$\bar{x} = r_1 m_1 + r_2 m_2 + \cdots r_k m_k$$

$$s = \sqrt{r_1 (m_1 - \bar{x})^2 + r_2 (m_2 - \bar{x})^2 + \cdots + r_k (m_k - \bar{x})^2}$$

Thus, Table 2.2 is augmented as shown in Table 2.3. The grouped means and standard deviations of Tables 2.2 and 2.3 are

$$\bar{x} = 0.04 \cdot 14.5 + 0.11 \cdot 24.5 + 0.27 \cdot 34.5 + 0.38 \cdot 44.5 + 0.11 \cdot 54.5$$
$$+ 0.07 \cdot 64.5 + 0.02 \cdot 74.5 = 41.5$$
$$s^2 = 0.04(14.5 - 41.5)^2 + 0.11(24.5 - 41.5)^2 + 0.27(34.5 - 41.5)^2 + 0.38(44.5 - 41.5)^2$$
$$+ .11(54.5 - 41.5)^2 + 0.07(64.5 - 41.5)^2 + 0.02(74.5 - 41.5)^2 \approx 155.00$$
$$s = \sqrt{155} \approx 12.4$$

Group index i	Age	Frequency f_i	Relative frequency r_i	Mark m_i
1	10–19	2	0.04	14.5
2	20–29	5	0.11	24.5
3	30–39	12	0.27	34.5
4	40–49	17	0.38	44.5
5	50–59	5	0.11	54.5
6	60–69	3	0.07	64.5
7	70–79	1	0.02	74.5
		45	1	

Table 2.3 Ages in an adult education class

EXERCISES 2.1

1. Find the standard deviation of the data {1, 3}.

2. Find the standard deviation of the data {1, 1, 1, 1, 3, 3, 3, 3}.

3. Find the standard deviation of the data {10, 21, 15, 14, 9, 12, 8, 10, 9, 11, 12 }.

4. Find the standard deviation of the data {23, 20, 31, 38, 26, 19, 35, 32, 27, 32, 25}.

5. Find the standard deviation of the data {2.2, 1.5, 1.7, 2.0, 0.9, 2.3, 1.3, 1.7, 0.8, 1.2, 1.7}.

6. a) Organize the following data by creating a frequency table:

   2 1 3 0 1 0 2 5 1 2 0

   2 2 1 4 3 1 1 0 0 0 2

   2 1 0 3 1 1 3 4 2 1 3

   0 1

   b) Compute the mean and the standard deviation of the data of part a.

7. a) Organize the following data by creating a frequency table:

   66 61 66 61 63 68 64 67 60 63 69

   67 67 60 67 60 62 68 65 60 64 65

   60 64 62 67 62 69 63 65 64 69 63

   69 64 68 67 61 68 69

   b) Compute the mean and the standard deviation of the data of part a.

8. a) Organize the following data by creating a frequency table with groups 0.5 – 0.9, 1.0 – 1.4, etc.:

   1.0   1.8   2.4   1.2   2.0   0.7   1.3   2.4   1.1   3.3   2.4

   0.8   2.3   1.7   1.0   2.8   1.4   3.0   0.8   1.2   1.4   2.2

   1.5   3.2   2.1   2.7   1.8   1.6   2.3   2.6   1.3   2.9   1.9

   1.2   0.5

   b) Compute the mean and the standard deviation of the data of part a.

9. a) Organize the following data by creating a frequency table with groups 5 – 9, 10 – 14, etc.

   24   32   24   32   11   30   10   11   28   20   11

   25   25   30   22   36   30   24   17   26   21   25

   29   37   31   39   19   25   22   8   37   29   19

   16   26   15   24   13   29   21   19   35   15   23

   29   24   27   26   20   18

   b) Compute the grouped mean and the standard deviation of the data of part a.

10. a) Complete the table below to a frequency table and draw its histogram:

Group	Frequency
0–9	99
10–19	273
20–29	196
30–39	79
40–49	67
50–59	20

b) Compute the grouped mean and the standard deviation of the data of part a.

11. a) Complete the table below to a frequency table and draw its histogram:

Group	Frequency
30–36	9
37–43	73
44–50	96
51–57	9
58–64	7
65–71	6

b) Compute the grouped mean and the standard deviation of the data of part a.

12. a) Complete the table below to a frequency table and draw its histogram:

Group	Frequency
15.3–15.5	9
15.6–15.8	27
15.9–16.1	19
16.2–16.4	7
16.5–16.9	6

b) Compute the grouped mean and the standard deviation of the data of part a.

13. a) Complete the table below to a frequency distribution and draw its histogram:

Group	Frequency
16–18	111
19–21	1,380
22–24	1,158
25–27	928
28–30	847
31–33	591
34–36	878

b) Compute the grouped mean and the standard deviation of the data of part a.

14. Toss a die 100 times and organize the outcomes in the form of a frequency table. Compute the mean and the standard deviation of these data.

15. Verify that expression (2.1) is indeed 0 for the data {1, 1, 1, 1, 3, 3, 3, 3}.

16. Verify that expression (2.1) is indeed 0 for the data {1, 2, 3, 6, 10}.

2.2   Totaling Measurement Errors

The scientific custom of taking multiple observations of the same quantity and then selecting a single estimate that best represents it has its origin the early part of the 16th century. By that time European mariners had stopped hugging the shore and taken to sailing for months in uncharted seas with no signposts or landmarks. The ship's location could be computed to some extent on the basis of the position of the moon relative to the fixed stars as well as the difference between true north and the compass needle. All this called for observations that were unfortunately often beset with errors. These errors could be due to the motion of the ship, the imprecision of the instruments, the fallibility of the observers, atmospheric disturbances, and many other causes. It was to protect against such errors that the custom of repeating one's observations many times arose. This led to the natural question of how to decide on the best possible estimate. Should one use the "middle" observation, commonly known as the *median*? Should one *average* by adding all the measurements and dividing by their number? Should outlying observations that lie far from all the others be disregarded as patently irrelevant? As erroneous conclusions regarding the position of the ship could, and often did, lead to disastrous consequences, this was an important question.

At first each observer had his own method for producing the best estimate, which method was rarely, if ever, explicitly written down. A case in point is provided by Johannes Kepler (1571–1630), one of the founders of modern astronomy who at one time, in trying to pinpoint the location of a certain star, chose the numerical value 24.55 to represent the four observations: 23.65, 27.62, 23.30, and 29.80. Because the average of these four numbers is 26.09 and their median is 25.64, it is clear that he had neither of these popular estimates in mind. It is not known what he did have in mind.

There are many recorded instances of such arbitrary choices by astronomers. Sad to say, it seems that quite frequently observers would simply choose such a number as best confirmed their preconceived notions. Eventually, the need for a standardized procedure backed by a mathematical theory of observational errors was recognized by the scientific community. The first to make an attempt at an explicit theory of errors was none other than Galileo Galilei in his epoch-making book, *Dialog Concerning the Two Chief World Systems—Ptolemaic and Copernican*, which appeared in 1632 and was suppressed by the Roman Catholic Church.

Galileo's discussion of the errors inherent in astronomical observations was summarized as follows [Hald, 1990]:

G1. *There is only one number which gives the distance of the star from the center of the earth, the true distance.*

G2. *All observations are encumbered with errors, due to the observer, the instruments, and the other observational conditions.*

G3. *The observations are distributed symmetrically about the true value; that is, errors are distributed symmetrically about zero.*

G4. *Small errors occur more frequently than large errors.*

G5. *The most probable hypothesis is the one which requires the smallest corrections of the observations.*

Points G1, G2, and G4 are self-explanatory. Point G3 is to be understood as referring to all *potential* measurements of a certain quantity rather than any actual set of such numbers. After all, it is quite conceivable that two measurements of some length should both be overestimates. An alternative statement of this principle says

G3′. *If e is any real number, then the errors of magnitudes e and −e are equally likely to occur.*

The discussion of Galileo's fifth, and most interesting, point requires a clarification of his terminology. Galileo thought of every measurement as being accompanied by an error and therefore needing, in principle, a correction. Multiple measurements resulted in multiple errors whose cumulative effect needed to be taken into consideration. But how should the errors be combined? It would clearly be misguided to simply add them and let overestimates cancel out underestimates, for the same reason that two hunters whose shots pass, respectively, one foot over and under the same duck must be conceded to have missed their mark. To prevent such fallacious cancellations, Galileo drew a slight distinction between an *error* and its *correction*. The latter is the positive part (technically, absolute value) of the former. Thus, both the errors −0.1″ and 0.1″ require the same correction of 0.1″.

Point G5 above is now illustrated by means of an example. Suppose a repeated measurement conducted by scientists A and B yielded the estimates 4.01, 4.03, and 4.11 (the units don't matter here) of the same quantity. Suppose further that scientist A prefers to think of 4.11 as an outlier, decides to ignore it, and proposes

$$4.02 = \frac{4.01 + 4.03}{2}$$

as an estimate of the true value. Scientist B, however, does not wish to throw away any of the data and proposes

$$4.05 = \frac{4.01 + 4.03 + 4.11}{3}$$

as his candidate for the true value. Which of these two estimates is more likely to be the true value? The import of Galileo's point G5 is that estimates should be judged by sum of the corrections they entail. For A's and B's estimates, these sums are, respectively,

4.02:
$$|4.01 - 4.02| + |4.03 - 4.02| + |4.11 - 4.02|$$
$$= 0.01 + 0.01 + 0.09 = 0.11$$

and

4.05:
$$|4.01 - 4.05| + |4.03 - 4.05| + |4.11 - 4.05|$$
$$= 0.04 + 0.02 + 0.06 = 0.12$$

Because $0.11 < 0.12$ it follows from Galileo's criterion that 4.02 is a better candidate for the unknown true value than 4.05. As it happens, this is the only sense in which Galileo used this total corrections criterion: to decide between subjectively selected candidates for the true value.

However, in the example under consideration there is a better candidate yet: the middle measurement 4.03. The sum of its corrections is

4.03:
$$|4.01 - 4.03| + |4.03 - 4.03| + |4.11 - 4.03|$$
$$= 0.02 + 0 + 0.08 = 0.10$$

The numbers below indicate that 4.03 is in fact the best candidate for the true value, at least in Galileo's sense of minimizing the sum of the corrections:

Candidate:	4.01	4.02	4.03	4.04	4.05	4.06	4.07	4.08
Total correction:	0.12	0.11	0.10	0.11	0.12	0.13	0.14	0.15

The optimality of 4.03 can also be displayed visually by means of a graph. If x denotes any estimate of the measured quantity, then the total of the concomitant corrections is

$$y = |4.01 - x| + |4.03 - x| + |4.11 - x|$$

A graph of this function, obtained by means of a calculating device, is displayed in Figure 2.3. The optimality of 4.03 manifests itself in the fact that the lowest point on this graph occurs at $x = 4.03$.

As noted above, this best candidate is the middle one of the three measurements. This turns out to be no coincidence. When the number of measurements is odd, the middle one, otherwise known as the *median*, is invariably the optimal Galilean candidate. When the number of measurements is even, either of the two middle ones or any number in between them will serve, and it is customary to use the

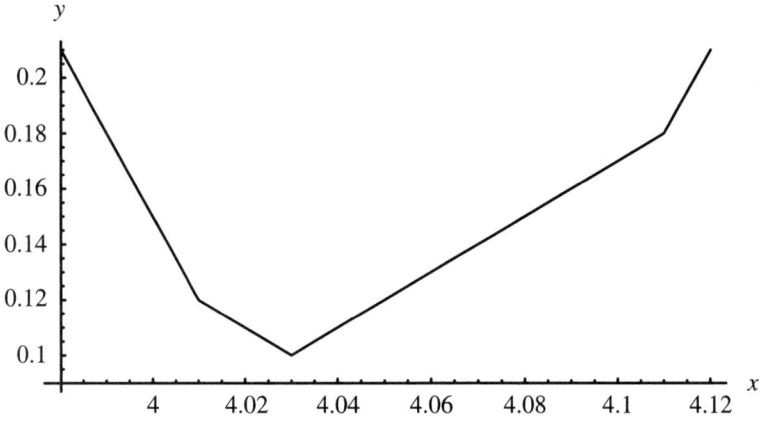

Figure 2.3 *The graph of a sum of corrections.*

midpoint of these two middle values as the *median*. This is illustrated in the table below, which lists the total corrections for a variety of candidates in the context of Kepler's four data mentioned above: 23.65, 27.62, 23.30, and 29.80.

Candidate:	23.50	24.00	25.00	26.00	27.00	28.00	29.00	29.50
Total correction:	10.77	10.47	10.47	10.47	10.47	11.23	13.23	14.23

Once again, the optimality of the median is also demonstrated visually by means of the graph in Figure 2.4.

Because Galileo did not comment on this interesting coincidence, it may be concluded that he did not notice it, probably because he expected the characteriza-

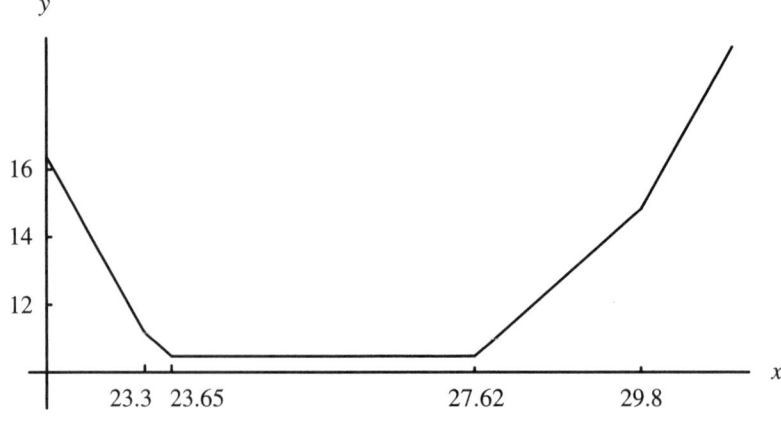

Figure 2.4 *The graph of a sum of corrections.*

tion of the best candidates to require a much more complicated resolution and so chose not to pursue it. He was, after all, preoccupied with the true nature of the solar system, and at that time not much depended on the question of what to do with discrepant multiple observations.

Galileo's fifth point, proposing the minimization of the total error, can be viewed as an instance of *Occam's razor* at work. This razor is an informal, though widely accepted, philosophical/scientific principle which stipulates that

> All else remaining the same, the simplest possible hypothesis should be accepted as the most likely explanation.

In the context of multiple errors, the *simplest* possible hypothesis could be interpreted as the estimate that calls for the *least* amount of correction.

Galileo's suggestion that the best candidate is that which minimizes the corrections was by and large, ignored by other astronomers and mathematicians. Instead, a variety of mostly undocumented ad hoc procedures were used. Eventually, sometime during the 18th century, the average of the measurements became widely accepted as the most likely true value. It is not known how this preference became established. The justification offered by Roger Cotes (1682–1716), one of Sir Isaac Newton's disciples, for this choice was that if the various measurements $m_1, m_2, ..., m_n$ were pictured on the number line and equal weights were placed at these points (Fig. 2.5), then the center of gravity of this system of weights would be located at exactly

$$\bar{m} = \frac{m_1 + m_2 + \cdots + m_n}{n}$$

which is of course the average of the measurements. This explanation was not universally accepted, and scholarly arguments against the use of the average can still be found as late as 1777. After all, measurements are not corporeal bodies and do not possess weight. Why then should this fictitious center of gravity of erroneous measurements have anything to do with the most likely true value?

At the beginning of the 19th century, two of the foremost mathematicians of the time, the German C. F. Gauss (1777–1855) and the Frenchman A. M. Legendre (1752–1833), proposed, independently of each other, an optimality criterion that bears the same relation to the average as Galileo's least total corrections principle bears to the median. This criterion, known as the *least squares method*, called for

Figure 2.5 *Balancing measurements.*

squaring the errors rather than taking their absolute values before they are totaled. For example, in reference to the data 4.01, 4.03, and 4.11 used above, the totaled squared corrections are as follows:

<u>4.02</u>:
$$(4.01 - 4.02)^2 + (4.03 - 4.02)^2 + (4.11 - 4.02)^2$$
$$= (-0.01)^2 + 0.01^2 + 0.09^2 = 0.0083$$

<u>4.03</u>:
$$(4.01 - 4.03)^2 + (4.03 - 4.03)^2 + (4.11 - 4.03)^2$$
$$= (-0.02)^2 + 0^2 + 0.08^2 = 0.0068$$

<u>4.05</u>:
$$(4.01 - 4.05)^2 + (4.03 - 4.05)^2 + (4.11 - 4.05)^2$$
$$= (-0.04)^2 + (-0.02)^2 + 0.06^2 = 0.0056$$

<u>4.06</u>:
$$(4.01 - 4.06)^2 + (4.03 - 4.06)^2 + (4.11 - 4.06)^2$$
$$= (-0.05)^2 + (-0.03)^2 + 0.05^2 = 0.0059$$

Of the above sums of squares, the least is 0.0056, given by 4.05, which also happens to be the average of 4.01, 4.03, and 4.11. The optimality of 4.05 can also be displayed visually by means of a graph. If x denotes any estimate of the measured quantity, then the total of the squares of the concomitant errors is

$$y = (4.01 - x)^2 + (4.03 - x)^2 + (4.11 - x)^2$$

A graph of this function, obtained by means of a calculating device, is displayed in Figure 2.6. This graph's lowest point lies right over $x = 4.05$.

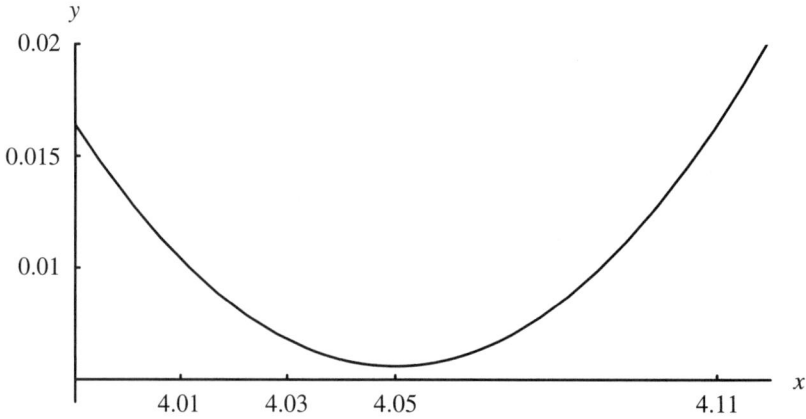

Figure 2.6   *The graph of a sum of squares of corrections.*

Let us reexamine Kepler's measurements 23.65, 27.62, 23.30, 29.80 from this point of view. The sums of the squares of the corrections for the listed candidates are as follows:

<u>24</u>: $(23.65 - 24)^2 + (27.62 - 24)^2 + (23.30 - 24)^2 + (29.80 - 24)^2 = 47.36$

<u>25</u>: $(23.65 - 25)^2 + (27.62 - 25)^2 + (23.30 - 25)^2 + (29.80 - 25)^2 = 34.62$

<u>25.635</u> (median): $(23.65 - 25.635)^2 + (27.62 - 25.635)^2 + (23.30 - 25.635)^2$
$+ (29.80 - 25.635)^2 = 30.68$

<u>26.092</u> (average): $(23.65 - 26.092)^2 + (27.62 - 26.092)^2 + (23.30 - 26.092)^2$
$+ (29.80 - 26.092)^2 = 29.84$

<u>27</u>: $(23.65 - 27)^2 + (27.62 - 27)^2 + (23.30 - 27)^2 + (29.80 - 27)^2 = 33.14$

Of these candidates, 25.635 is the median of Kepler's data and 26.092 is the average. Once again it is the average that yields the minimum sum of the squares of the corrections. The optimality of the average is also demonstrated by Figure 2.7, which displays the graph of the sum of the squared errors:

$$y = (23.65 - x)^2 + (27.62 - x)^2 + (23.30 - x)^2 + (29.80 - x)^2$$

Here too the graph's lowest point lies over the average $x = 26.092$.

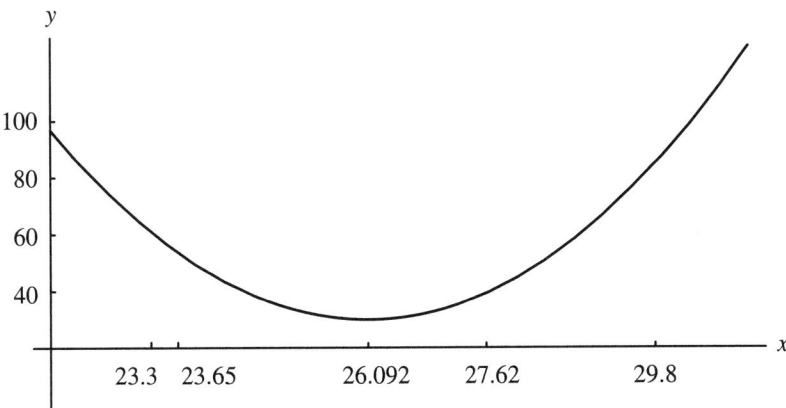

Figure 2.7   *The graph of a sum of squares of corrections.*

EXERCISES 2.2

For each of the sets of measurements in Exercises 1–8,

a) Compute the median and the average.

b) Compute the sum of the corrections associated with both the median and the average. Which is smaller?

c) Compute the sum of the squares of the errors associated with both the median and the average. Which is smaller?

   1. {1, 2, 3, 4, 10}   2. {9, 9, 9, 8, 7, 6, 6, 6, 5, 5}

   3. {5.1, 5.2, 5.3, 5.4, 6.0}   4. {10.2, 10.2, 10.3, 10.3, 10.5}

   5. {9.1, 9.1, 9.1, 9.2, 9.2}   6. {1.5, 1.6, 1.8, 1.9, 2.0, 2.0}

   7. {8.5, 8.5, 8.5, 8.7, 8.7, 8.7}   8. {8.5, 8.7, 8.7, 8.7, 8.7, 8.7}

2.3   Least Squares and Regression

The context for Legendre's proposal of the least squares was that of geodesy—the scientific discipline concerned with the exact shape of the earth. In the spirit of their country's recent revolution, France's scientists had decided to abandon their antiquated measurement system and to adopt a new one, more scientific and efficient, whose unit of length, the *meter*, would equal

$$\frac{1}{40,000,000}$$

of the circumference of the earth. This necessitated an accurate determination of the said circumference that, in turn, depended on the exact shape of the earth. They were aware that this shape deviated somewhat from the perfect sphere. This distortion is due to the centrifugal force produced by the earth's daily rotation around its axis, which slightly flattens its north–south silhouette into an ellipse, i.e., an oval shape whose horizontal axis is somewhat longer than its vertical axis. To determine the exact dimensions of these axes, observations were taken in 1799 along a nearly 700-mile-long arc that, of course, lay completely within the French territories of the time. As indicated by Figure 2.8, ellipses come in many shapes, and the question then arose of which ellipse best fitted these observations. It was to answer this question that the least squares approach was adopted.

    A different procedure, one based on the Galilean approach of totaling the corrections and known as the Boscovich method, had been proposed in 1770. It never gained popularity and sank into virtual oblivion once the least squares approach had

Figure 2.8 *A variety of ellipses.*

been promulgated. The reasons for the scientists' preferences for the latter method may seem strangely subjective and possibly irrelevant. The mathematics of Legendre's method was *easier* and more *satisfactory*. Note that the curves of Figures 2.3 and 2.4, which display the sum of the *corrections*, have corners, whereas those of Figures 2.6 and 2.7, which display the sums of the *squares of the errors*, are smooth, albeit curved. This difference means that the methods of calculus, an important mathematical discipline whose development was intimately tied to the evolution of modern science and technology, are applicable only to the latter but not to the former. It was this applicability of calculus to the method of least squares that made the scientific community prefer it over the Boscovich method. In other words, it was the amenability of the method of least squares to the tools of calculus rather than any objective demonstration of its superiority that drew mathematicians to it. Considering modern science's insistence on the objectivity of its theories and on the need to verify them by means of factual observation, this subjective criterion of convenience is rather surprising. Another reason for the alacrity with which the scientific community adopted the least squares approach may have been that it was so closely associated with the averaging procedure that, by some folk process whose details are lost, had already been adopted by the same group during the preceding half century.

We now show how the method of least squares is used to make predictions on the basis of experience and/or observations.

Example 2.3.1. *A retail merchant's experience indicates a fairly direct correlation between his sales and his advertising expenditures. In particular, the following data were recorded on 4 consecutive weeks:*

Advertising costs ($):	40	20	30	25	(2.3)
Sales ($):	4850	3150	4500	4050	

Using these data, the merchant wishes to describe the relationship between the expenditures and the sales as a function $y = f(x)$, where x denotes the expenditures and y the sales. Such a function would make it possible for him to project the sales resulting from any amount he might wish (or have available) to spend on advertising. Because the relationship between advertising and sales is subject to random fluctuations, this projection would not be accurate but would still be better than no projection at all. Such functions and equations that represent the way real world variables affect each other are called models.

   Having no other information to go on, the merchant should assume that the function $f(x)$ is as simple as possible. This means that $f(x)$ is *linear* in the sense that its graph is a straight line or, more technically, there are constants m and b such that

$$f(x) = mx + b$$

The specific values of m and b are then to be determined from the data of (2.3). Had there been only two observations [say the first two values in (2.3)], the procedure would have been straightforward. One would interpret these observations as the points (40, 4850), (20, 3150) of the Cartesian coordinate plane and then look for the equation of the straight line that passes through these two points. The two-point form of the equation of this straight line is

$$y - 3150 = \frac{4850 - 3150}{40 - 20}(x - 20) = 85(x - 20) = 85x - 1700$$

or

$$y = 85x - 1700 + 3150 = 85x + 1450$$

This function is demonstrably inaccurate at $x = 25$ and $x = 30$. At these values it predicts sales of

$$f(25) = 85 \cdot 25 + 1450 = 3575 \quad \text{and} \quad f(30) = 85 \cdot 30 + 1450 = 4000$$

respectively, which differ from the actual 4050 and 4500. These discrepancies are illustrated in Figure 2.9.

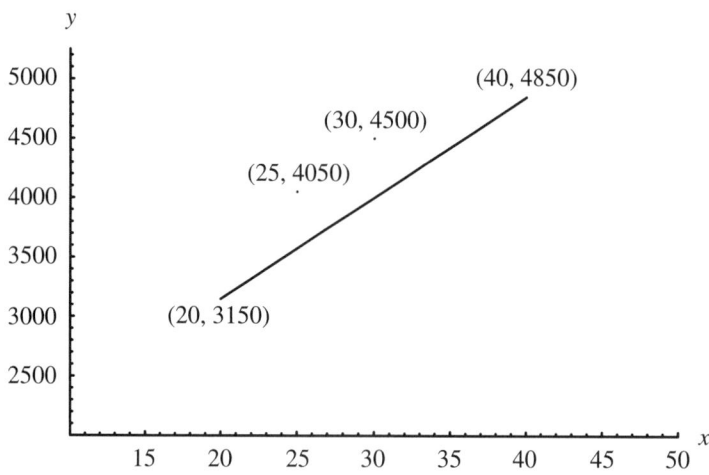

Figure 2.9   *An attempt to fit a line to four given points.*

Such errors are endemic to the situation. Because of the inherent randomness of the universe, data points rarely align themselves along a perfectly straight line, and so any proposed straight line model will entail some errors. If the data points are $(x_1, y_1), (x_2, y_2), ..., (x_n, y_n)$ and the model is $f(x) = mx + b$, then the errors are defined as

$$e_i = y_i - (mx_i + b) \quad i = 1, 2, ..., n$$

For example, for the data points of Table (2.3) and the model $y = 90x + 1300$, the errors are:

$$e_1 = 4850 - (90 \cdot 40 + 1500) = -250 \quad e_2 = 3150 - (90 \cdot 20 + 1500) = -150$$
$$e_3 = 4500 - (90 \cdot 30 + 1500) = 300 \quad e_4 = 4050 - (90 \cdot 25 + 1500) = 300$$

Because the errors are inevitable, they can be at best minimized, and it is commonplace to use the least squares method criterion proposed by Gauss and Legendre. In other words, we make the following assumption:

That model is best which minimizes the sum of the squares of the errors.

The straight line $y = mx + b$ that minimizes the sum of the squares of the errors is called the *regression* line. Its parameters m and b are computed by the following procedure:

For the data points $(x_1, y_1), (x_2, y_2), ..., (x_n, y_n)$ set:

$$S_x = x_1 + x_2 + \cdots + x_n \qquad S_y = y_1 + y_2 + \cdots + y_n$$
$$S_{x^2} = x_1^2 + x_2^2 + \cdots + x_n^2 \qquad S_{xy} = x_1 y_1 + x_2 y_2 + \cdots + x_n y_n$$

Then,

$$m = -\frac{nS_{xy} - S_x S_y}{nS_{x^2} - (S_x)^2} \quad \text{and} \quad b = \frac{S_y - mS_x}{n}$$

For example, using the four data points of (2.3), we obtain

$$S_x = 40 + 20 + 30 + 25 = 115$$
$$S_y = 4850 + 3150 + 4500 + 4050 = 16,550$$
$$S_{x^2} = 40^2 + 20^2 + 30^2 + 25^2 = 3525$$
$$S_{xy} = 40 \cdot 4850 + 20 \cdot 3150 + 30 \cdot 4500 + 25 \cdot 4050 = 493,250$$
$$m = \frac{4 \cdot 493,250 - 115 \cdot 16,550}{4 \cdot 3525 - 115^2} \approx 79.7$$
$$b = \frac{16,550 - 79.7 \cdot 115}{4} \approx 1846.1$$

These calculations yield the regression line $y = 79.7x + 1846.1$. Using this regression line, one would project that if the merchant of table (2.3) were to spend $60 on advertising, his sales would go up to

$$79.7 \cdot 60 + 1846.1 = 6628.1 \text{ dollars.}$$

It might also be of interest to compute this model's errors:

$$e_1 = 4850 - (79.7 \cdot 40 + 1846.1) \approx -184 \quad e_2 = 3150 - (79.7 \cdot 20 + 1846.1) \approx -290$$
$$e_3 = 4500 - (79.7 \cdot 30 + 1846.1) \approx 263 \quad e_4 = 4050 - (79.7 \cdot 25 + 1846.1) \approx 211$$

This regression line and its errors are illustrated in Figure 2.10.

The credibility of the average and the method of least squares were greatly enhanced by the *Ceres* incident. On January 1, 1801 the Italian astronomer Giuseppe Piazzi sighted a heavenly body that he strongly suspected to be a new planet. He announced his discovery and named it *Ceres*. Unfortunately, 6 weeks later, before enough observations had been taken to make possible an accurate determination of its orbit, so as to ascertain that it was indeed a planet, Ceres disappeared behind the sun and was not expected to reemerge for nearly a year. Interest in this possibly new planet was widespread, and astronomers throughout Europe prepared themselves by compu-guessing the location where Ceres was most likely to reappear. The young Gauss, who had already made a name for himself as an extraordinary mathematician, proposed that an area of the sky be searched that was quite different from those suggested by the other astronomers; he turned out to be right. He became a celebrity, and the theory of observational errors that helped him locate Ceres was immediately recognized as a milestone in the evolution of the scientific method.

Gauss' theory of errors was based on the following three assumptions:

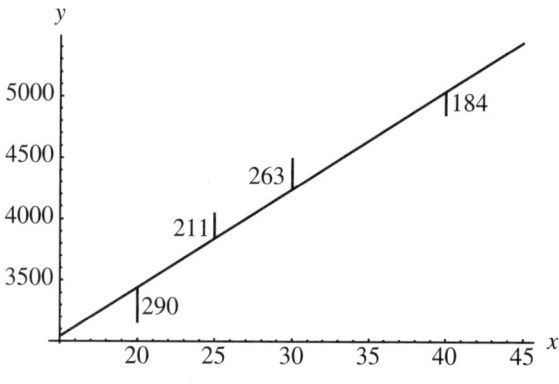

Figure 2.10 A regression line.

N1.   *Small errors are more likely than large errors.*

N2.   *For any real number e the likelihoods of errors of magnitudes e and -e are equal.*

N3.   *In the presence of several measurements of the same quantity, the most likely value of the quantity being measured is their average.*

On the basis of these principles and some clever mathematics, Gauss then concluded that the mathematical curves that best described the distribution of observational errors were none other than the normal curves that were being used for approximating binomial probabilities. In other words, Gauss proved that the histograms of data obtained by taking multiple measurements of the same object will, in general, have the shape of the normal curve. Figure 2.11 displays some astronomical measurements pertaining to the right ascension of the Polar Star that originate in the publications of the Royal Observatory of Greenwich. Because the unit of measurement is rather technical, it is omitted. After all, it is the shape of the data that is the main concern here.

Gauss's assumptions N1 and N2 stand to reason and had also been stated by Galileo. The third principle was revolutionary and inspired. It should be borne in mind that averaging had only become commonplace among astronomers during the previous half century, and even then its adoption was in all likelihood a matter of subjective convenience rather than objective science. Despite the ubiquity of averaging, there was no proof, either theoretical or experimental, that the average

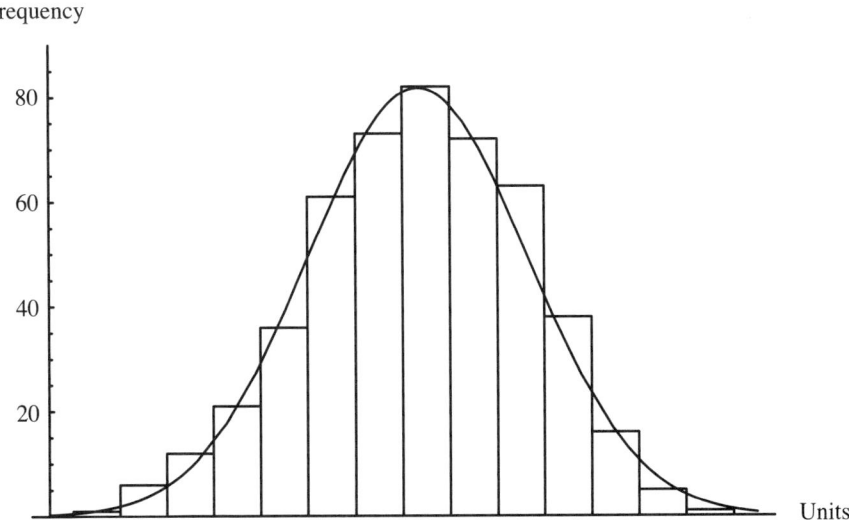

Figure 2.11   *The histogram of some astronomical measurements.*

constitutes a better candidate for the true value than the median or any other estimator. Nor has such a proof been offered since then.

This is a rather interesting issue. The normal curve, which was devised as a tool for approximating binomial probabilities and then was used by Gauss to describe the distribution of measurement errors, was later demonstrated to have a myriad of other applications to the real world (see Sections 2.4 and 2.5). So much so that it is now considered to be the cornerstone of both mathematical and applied statistics. Nevertheless, the justification of the eminent position it holds in science boils down to the *subjective* criterion that averaging is the best way to handle the difficulties presented by random errors.

Gauss himself seems to have been uncomfortable with Assumption N3, though not with the conclusion that the error curves are normal. He subsequently justified the normality of the error curve by means of a variation of the least squares approach described above. However, as was noted before, the preference for least squares is also based on highly subjective grounds.

The reason for the applicability of De Moivre's approximation curve to the distribution of errors has been explained as follows: Assume that just like matter, errors have an atomic structure so that each error is actually an aggregate of "atomic" or elementary errors, all of which have equal magnitude. If these elementary errors occur independently of each other and with equal probabilities, then one can think of the probability of occurrence of an actual error as a binomial probability whose binomial trials are the elementary errors. We hasten to add that this is hardly a scientific explanation. The existence of elementary errors was hypothesized solely for the purposes of this and other explanations. They have no factual basis.

This section is brought to a close with a restatement of the three *distinct but equivalent* scientific principles that can be used to explain the central role played by the normal curve in the handling of random errors:

R1.   *The average is the best estimate for the true value.*

R2.   *When aggregating errors, the individual errors should be squared.*

R3.   *The distribution of measurement errors is best described by the normal curve.*

EXERCISES 2.3

For each of the sets of data points in Exercises 1–8 below,

   a)   Find the line of regression.

   b)   Compute all the errors associated with the line of regression.

   c)   Use the line of regression to estimate the value of y that corresponds to x_0.

d) Graph the points and the regression line as well as its associated errors.

e) Find the equation of the straight line that contains the first and last points. Compute the errors associated with this line and verify that the sum of their squares is at least as large as the corresponding sum for the regression line of part a.

1. $\{(0, 2), (1, 1)\}$, $x_0 = 2$

2. $\{(0, 0), (1, 2)\}$, $x_0 = 2$

3. $\{(0, 0), (1, 1), (2, 2)\}$, $x_0 = 3$

4. $\{(0, 0), (1, 1), (2, 3)\}$, $x_0 = 5$

5. $\{(1, 2), (2, 3), (3, 2)\}$, $x_0 = 0$

6. $\{(0, 5), (1, 4), (3, 3), (4, 1)\}$, $x_0 = 2$

7. $\{(10, 26), (15, 20), (25, 10)\}$, $x_0 = 20$

8. $\{(12, 55), (18, 30), (20, 25)\}$, $x_0 = 10$

9. The following data are the selling prices y of a certain make and model of used car x years old:

x (years):	1	2	3	5
y (dollars):	6250	5725	5350	4945

Find the regression line for these data and use it to project the price of a 10-year-old car.

10. The grades of a class of five students on a midterm exam (x) and on the final exam (y) are as follows:

x (midterm grade):	77	50	71	81	99
y (final grade):	82	66	78	47	99

Find the regression line of these data and use it to predict the final exam grade of a student who scores 60 on the midterm.

11. To plan for contingencies, a municipal utility wishes to determine the relationship between the average temperature (in degrees Fahrenheit) and the average daily household consumption of water (in gallons). Given the following data:

	January	April	July	October
Average daily temperature	26	46	71	61
Average daily consumption	180	193	245	238

find the line of regression of these data and use it to project the average daily water consumption in a month whose average daily temperature is 50°F.

2.4 Normally Distributed Data

As the saying goes, nothing succeeds like success, and subsequent to Gauss's spectacular prediction of the location of Ceres, the least squares approach as well as the normality of the distribution of random errors soon became standard tools of all the physical sciences. It took another 50 years for scientists to come to the realization that they had in their possession a tool whose scope far exceeded the mere description of measurement errors. This next step was taken by the Belgian astronomer and sociologist Adolphe Quetelet (1796–1874), who was struck by the possibility that most data, and not just measurements and their accompanying errors, when grouped into classes might display a normal shape. His first major example, given in 1846, was drawn from an 1817 article of the *Edinburgh Medical and Surgical Journal* and listed the frequencies of chest measurements, in inches, of 5738 Scottish soldiers:

Chest measurements of Scottish soldiers

Girth	Frequency
33	3
34	18
35	81
36	185
37	420
38	749
39	1073
40	1079
41	934
42	658
43	370
44	92
45	50
46	21
47	4
48	1
	5738

These data are graphed in Figure 2.12 together with a normal curve that follows the general contours of the histogram. Quetelet did not justify his claim of the normality of the data by means of such a drawing. Instead, he asserted that this table

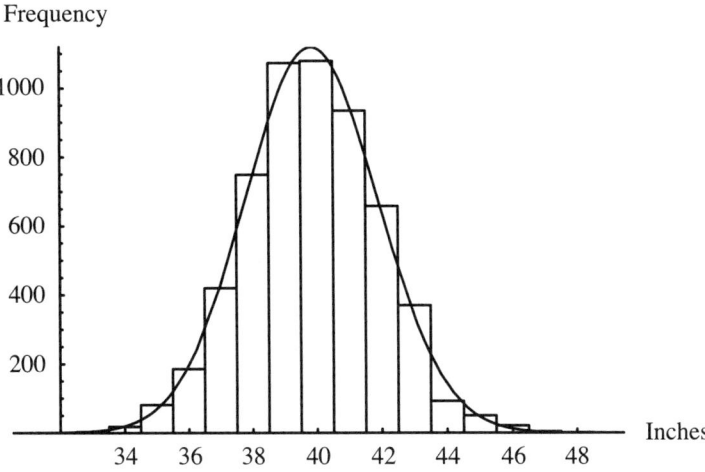

Figure 2.12   *Chest measurements of Scottish soldiers.*

would be practically indistinguishable from a table produced by an inexperienced person taking 5738 measurements of the chest of some statue. Given the subsequent widespread belief in the normality of all data and the historical role that this example played in the creation of this tenet, Quetelet's own words on this topic should be of interest:

> I now ask if it would be exaggerating, to make an even wager that a person little practiced in measuring the human body would make a mistake of an inch in measuring a chest of more than 40 inches in circumference? Well, admitting this probable error, 5,738 measurements made on one individual would certainly not group themselves with more regularity, as to the order of magnitude, than the 5,738 measurements made on the scotch soldiers; and if the two series were given to us without their being particularly designated, we should be much embarrassed to state which series was taken from 5,738 different soldiers, and which was obtained from one individual with less skill and ruder means of appreciation.
>
> *Letters Addressed to H. R. H. the Grand Duke of Saxe Coburg and Gotha, on the Theory of Probabilities, As Applied to the Moral and Political Sciences,* Translated by O. G. Downes, Leyton, London, 1849, p. 92.

In other words, Quetelet suggested that the deviations of the chest measurements should be viewed as errors committed by an imprecise natural process. And because measurement errors were assumed by astronomers (of which Quetelet was one) to be governed by the normal curve, it would be reasonable to expect the chest measurements to also fall into the same pattern.

Quetelet's assertion that the chest measurements have a normal distribution should be taken with a healthy dose of skepticism, particularly so in view of the fact that he made several arithmetical mistakes in transcribing his data. The grounding of the belief in this supposedly universal "shape" of data is not unshakable and has been attacked by some specialists. Nevertheless, in the absence of compelling reasons for believing otherwise, it is quite customary to assume the normality of most data. This assumption has very strong computational consequences that are very similar to those that proved so useful in the computation of binomial probabilities but have a much wider range of applications. These are the subject of this section's remainder.

It is common practice in statistics to use the term *population* to refer to any collection of data. If $\{x_1, x_2, ..., x_n\}$ is a population, then F_T denotes the *cumulative frequency* (i.e., number) of the data that do not exceed the threshold T and

$$R_T = -\frac{F_T}{n}$$

denotes their *cumulative relative frequency*. For example, for the chest measurements of the Scottish soldiers, we have

$$F_{36} = 3 + 18 + 81 + 185 = 287$$

$$R_{36} = \frac{287}{5738} \approx 0.05$$

and

$$F_{40} = 3 + 18 + 81 + 185 + 420 + 749 + 1073 + 1079 = 3608$$

$$R_{40} = \frac{3608}{5738} \approx 0.63$$

The assumption of the *normality of a population* is tantamount to the following statement.

NORMALITY ASSUMPTION 2.4.1. *If a population has average $\bar{x}$ and standard deviation s, then*

$$R_T \approx A\left(\frac{T - \bar{x}}{s}\right) \quad \text{and} \quad F_T \approx n \cdot A\left(\frac{T - \bar{x}}{s}\right)$$

---

where $A(t)$ denotes the entry of Table A that corresponds to a t that is described in the left and top borders of the table. When the possible values of the threshold T are restricted to relatively few integers, it is customary to replace the numerator $T - x$ by $T + 0.5 - x$.

As a first example we consider the ages of the last example of Section 2.1 for which the grouped mean and standard deviation were computed to be

$$\bar{x} = 41.5 \quad \text{and} \quad s = 12.4$$

Accordingly, the number of students of at most age 49 is

$$F_{49} \approx 45 \cdot A\left(\frac{49.5 - 41.5}{12.4}\right) = 45 \cdot A(0.65) = 45 \cdot 0.7422 \approx 33$$

Note that the actual number of students in this age group is

$$2 + 5 + 12 + 17 = 36$$

Similarly, the number of students of at least age 40 is

$$45 - 45 \cdot A\left(\frac{39.5 - 41.5}{12.4}\right) = 45 - 45 \cdot A(-0.16) = 45 - 45 \cdot 0.4364 \approx 25$$

This is close to the actual number of

$$17 + 5 + 3 + 1 = 26$$

It should also be of interest to see to what extent Quetelet's chest measurements fall into the normal pattern. For this, the mean and standard deviation are

$$\bar{x} = 39.8 \quad \text{and} \quad s = 2.0$$

Consequently, in predicting the number of soldiers with chest measurements of 40 inches, the Normality Assumption 2.4.1 yields

$$f_{40} = F_{40} - F_{39} \approx 5738 \cdot A\left(\frac{40.5 - 39.8}{2}\right) - 5738 \cdot A\left(\frac{39.5 - 39.8}{2}\right)$$
$$= 5738 \cdot A(0.35) - 5738 \cdot A(-0.15) = 5738 \cdot 0.6368 - 5738 \cdot 0.4404 \approx 1127$$

whereas the actual number is 1079.

Table 2.4 augments the Scottish soldiers' chest measurements table with the theoretical frequencies that were computed on the basis of the Normality Assumption 2.4.1. There are obvious disagreements between the actual and the theoretical (normally based) frequencies. Nevertheless, the fit between the two is close enough to have convinced social scientists of the validity of Quetelet's assertion that these and all similar data are normally distributed. This validity is still widely accepted by default whenever there is no compelling reason for rejecting it. Two examples of non-normally distributed data are offered here as illustrations. Data obtained by rolling a die 1000 times and recording the frequencies with which each face occurs will clearly *not* have a bell-shaped histogram and so are not normally distributed. Instead, the different frequencies will tend to remain equal to each other and their distribution is commonly described as uniform (Fig. 2.13). Similarly, suppose each of the students in a large class keeps tossing a coin until a *heads* shows up and records the number of

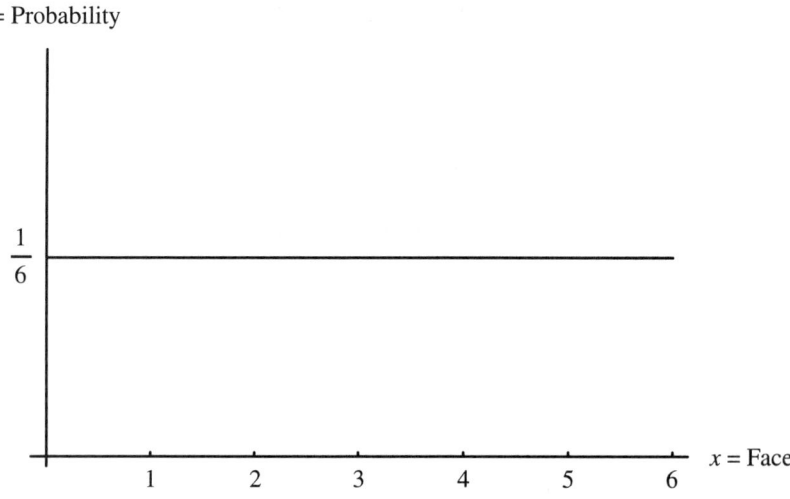

Figure 2.13   *A non-normal distribution that describes data generated by a fair die.*

Girth	Actual frequency	Predicted frequency
33	3	4
34	18	18
35	81	67
36	185	193
37	420	434
38	749	762
39	1073	1047
40	1079	1127
41	934	950
42	658	626
43	370	323
44	92	131
45	50	41
46	21	10
47	4	2
48	1	0
	5,738	

Table 2.4   Chest measurements of Scottish soldiers.

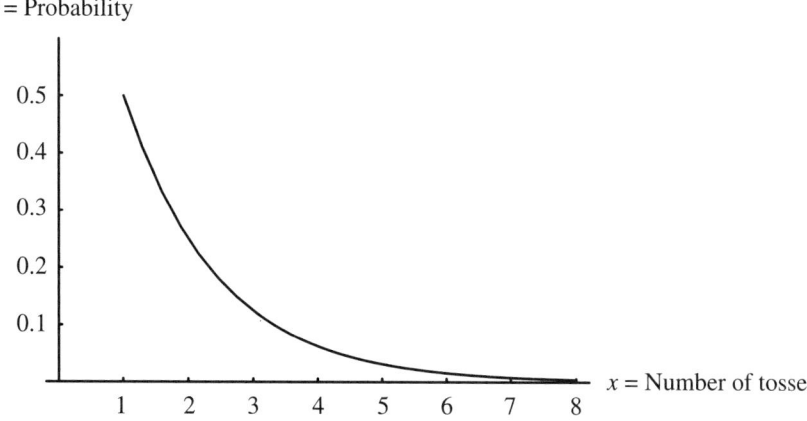

Figure 2.14   *A non-normal distribution that describes the number of coin tosses required to obtain a* heads.

tosses. One would then expect $\frac{1}{2}$ of the data to be 1, $\frac{1}{4}$ of the data to be 2, $\frac{1}{8}$ of the data to be 3, and so on. This is definitely a non-normal distribution. The histogram of this data would have the shape displayed in Figure 2.14.

Example 2.4.2. *An airline reports that the mileage recorded by frequent flyers averages 22,987 with a standard deviation of 6736. Assuming the normality of the data, estimate the proportion of flyers who have recorded the following mileages.*

a)   At most 20,000: $F_{20,000} \approx A\left(\dfrac{20,000 - 22,987}{6736}\right) \approx A(-0.44) = 0.33 = 33\%$

b)   At least 30,000: $1 - F_{30,000} \approx 1 - A\left(\dfrac{29,999 - 22,987}{6736}\right) \approx 1 - A(1.04)$

$$= 1 - 0.8508 = 0.1492 \approx 14.9\%$$

c)   Between 15,000 and 25,000: $F_{25,000} - F_{15,000}$

$$\approx A\left(\dfrac{25,000 - 22,987}{6736}\right) - A\left(\dfrac{14,999 - 22,987}{6736}\right) \approx A(0.30) - A(-1.19)$$

$$= 0.6179 - 0.1170 = 0.5009 \approx 50.1\%$$

Example 2.4.3. *Suppose next that the above airline wishes to award its top 5% customers a bonus advance of $100 toward their next ticket purchase and the next 10% an advance of $50. What are the cutoff mileages that qualify customers for these awards?*

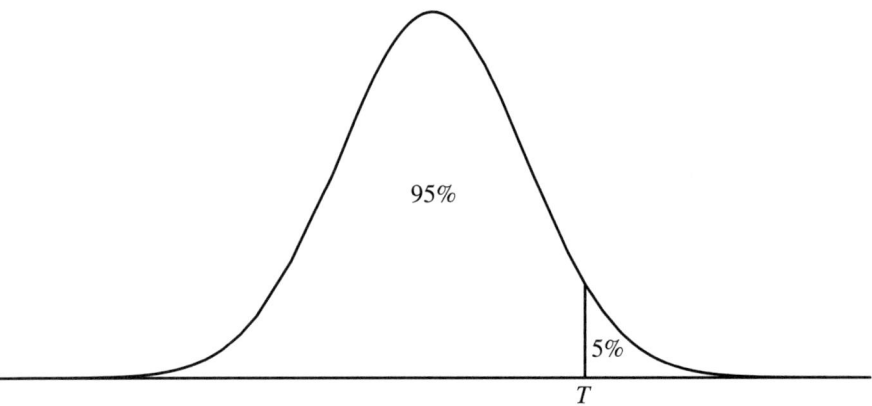

Figure 2.15 A 95% threshold.

Top 5%: We are looking for the threshold T below which 95% of the mileages lie (Fig. 2.15). Hence,

$$A\left(\frac{T - 22,987}{6736}\right) = 0.95$$

The entries of Table A that are closest to 0.95 are 0.9495 and 0.9505, and the latter is selected arbitrarily. Hence,

$$\frac{T - 22,987}{6736} = 1.65$$

By the rules of algebra we obtain the cutoff

$$T = 6736 \cdot 1.65 + 22,987 = 34,101.4$$

Thus, the company should award the $100 bonus to any customer who has logged at least 34,101 miles.

Next 10%: By the previous calculations, the top cutoff of this group is 34,101 miles. The bottom cutoff is the threshold T below which 85% of the data lie (Fig. 2.16). Hence,

$$A\left(\frac{T - 22,987}{6736}\right) = 0.85$$

The entry of Table A closest to 0.85 is 0.8508 and so

$$\frac{T - 22,987}{6736} = 1.04$$

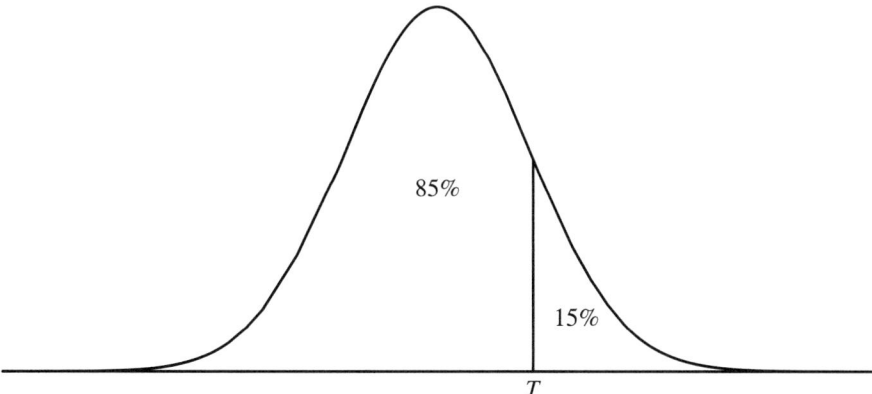

Figure 2.16 *An 85% threshold.*

By the rules of algebra we obtain the cutoff

$$T = 6736 \cdot 1.04 + 22{,}987 = 29{,}992.44$$

Thus, the $50 bonus should be awarded to all the customers with mileages between 29,993 and 34,101 inclusive.

For normal populations, it is a consequence of the Normality Assumption that the standard deviation provides a natural unit for describing the spread of the data. In particular, regardless of the values of the mean $\bar{x}$ and the standard deviation s,

34% of the data lie in each of the intervals $[\bar{x}, \bar{x} + s]$ *and* $[\bar{x} - s, \bar{x}]$,

14% of the data lie in each of the intervals $[\bar{x} + s, \bar{x} + 2s]$ *and* $[\bar{x} - 2s, \bar{x} - s]$,

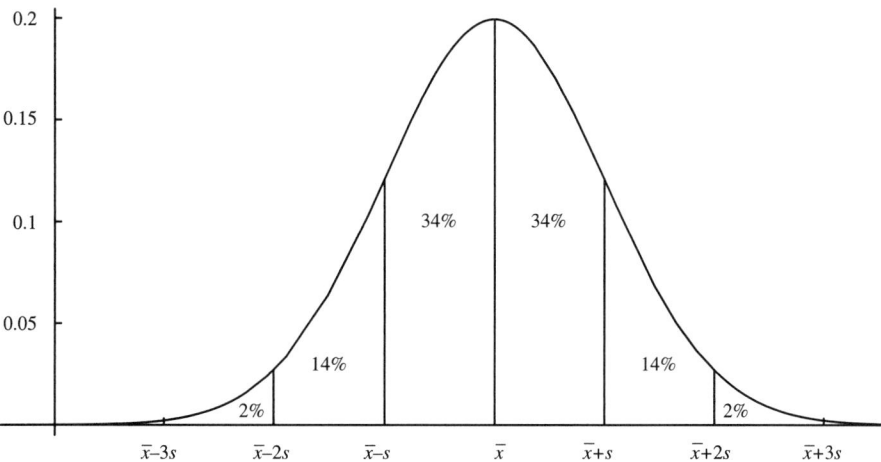

Figure 2.17 *Some standard distributions of data.*

2% of the data lie in each of the intervals $[\bar{x} + 2s, \bar{x} + 3s]$ and $[\bar{x} - 3s, \bar{x} - 2s]$

These percentages are illustrated in Figure 2.17. For example, because the mean girth among the Scottish soldiers is 39.8 inches and the standard deviation is 2 inches, one would expect 34% of 5738 (i.e., 1951) to have chests measuring either 40 or 41 inches. The actual number is

$$f_{40} + f_{41} = 1079 + 934 = 2014$$

Similarly, assume that the ages in the adult education class discussed in Section 2.1 are normally distributed. Because they are known to have mean $\bar{x} = 41.5$ and standard deviation $s = 12.4$, one would expect 14% of 45 = 6 of the students to have ages 54 – 66, which turns out to be exactly the right number.

A *random variable* is a phenomenon or a process that produces data. The daily high temperature in a certain city, the number of people in the next car to arrive at a toll booth, the amount of coffee dispensed by a vending machine into a cup, and the number of coin tosses required to accumulate 10 heads are all such random variables. Problems regarding the frequencies of certain types of data translate into questions regarding the probability of a random variable assuming corresponding values. Thus, the question of how many of the next 1000 cars passing through a toll booth contain at most two riders can be rephrased as finding the probability that such a car contains two riders.

Random variables also have means and standard deviations that are idealizations of the averages and standard deviations of the populations they produce. A random variable with mean 15 will produce populations with averages that are close to 15. A random variable with standard deviation 4 will produce populations with standard deviations that are close to 4. The Greek letters μ and σ are used to denote the mean and standard deviation of a random variable respectively. They are used much like the $\bar{x}$ and s of the data these variables generate.

A random variable is said to be normally distributed if the data it generates is normally distributed. If V denotes such a normally distributed random variable and T is any threshold, then the probability that V generates a number of value at most T is denoted by Prob[$V \leq T$] and is computed by the following analog of the Normality Assumption:

$$\text{Prob}[V \leq T] = A\left(\frac{T - \mu}{\sigma}\right) \tag{2.4}$$

For example, if in a certain town the high temperature on July 4 is a normally distributed random variable with mean $m = 94°F$ and standard deviation 5°F, then the probability that on the next July 4 the temperature will not exceed 100°F is

$$\text{Prob}[V \leq 100] = A\left(\frac{100 - 94}{5}\right) = A(1.2) = 0.8849 \approx 88.5\%$$

Example 2.4.4. *A vending machine is set to dispense 7 ounces of coffee into its cups. Due to the inevitable imperfections in its machinery, the amount that is actually poured out is a normally distributed random variable with a standard deviation of 0.2 ounces. To prevent unsightly spills, the machine is provided with cups that have volumes of 7.3 ounces.*

a) *What percentage of the dispensations will result in a spill anyway?* This is the probability that the amount poured exceeds 7.3, which, by Formula (2.4) equals

$$1 - A\left(\frac{7.3 - 7}{0.2}\right) = 1 - A(1.5) = 1 - 0.9332 \approx 6.7\%$$

b) *If too large a portion of the cup remains unfilled, the customer may feel shortchanged. Suppose the typical customer feels unhappy when half an ounce is left unfilled. What proportion of the customers will feel shortchanged?* Because 7.3 − 0.5 = 6.8, the proportion of unhappy customers is

$$A\left(\frac{6.8 - 7}{0.2}\right) = A(-1) = 0.1587 \approx 15.9\%$$

c) *To antagonize fewer customers, the vending company uses smaller cups. What should be the size of the cup if the company is willing to allow for spills to occur in 10% of the sales?* This is a threshold determination whose diagram appears in Figure 2.18. Hence,

$$A\left(\frac{T - 7}{0.2}\right) = 1 - 0.1 = 0.9$$

and, using Table A in the reverse direction,

$$\frac{T - 7}{0.2} = 1.28$$

or

$$T = 7 + 0.2 \cdot 1.28 = 7.256$$

Thus, a cup with capacity 7.256 ounces will reduce the number of unhappy customers while allowing for spills to occur no more than 10% of the time.

d) *What proportions of the drinks dispensed will contain between 6.7 and 7.2 ounces?* This is

$$A\left(\frac{7.2 - 7}{0.2}\right) - A\left(\frac{6.7 - 7}{0.2}\right) = A(1) - A(-1.5) = 0.8413 - 0.0668 \approx 77.5\%$$

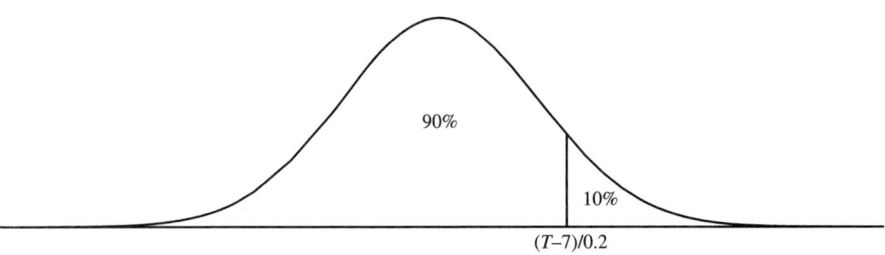

90%

10%

$(T–7)/0.2$

Figure 2.18   *A 90% threshold.*

EXERCISES 2.4

1. The yearly salaries of the employees of a certain large company are normally distributed with a mean of $28,500 and a standard deviation of $2800. What percentage of the employees falls in the following salary ranges:

    a) Between $20,000 and $30,000?   b)   More than $30,000?

    c) Between $25,000 and $35,000?   d)   Less than $32,000?

    e) A salary that deviates from the mean by more than $1000?

    f) A time that deviates from the mean by more than $4000?

2. The yearly salaries of the employees of a certain large company are normally distributed with a mean of $28,500 and a standard deviation of $2800. Find the cutoff salary for the following salary ranges:

    a) Top 20%   b)   Top 10%   c)   Bottom 15%

    d) Bottom 25%   e)   Top 40%   f)   Bottom 40%

3. The scores of a certain national test are normally distributed with a mean of 72 and a standard deviation of 7. Find the cutoff scores for the following groups of scores:

    a) Top 20%   b)   Top 10%   c)   Bottom 15%

    d) Bottom 25%   e)   Top 40%   f)   Bottom 40%

4. The scores of a certain national test are normally distributed with a mean of 72 and a standard deviation of 7. Find the percentage of scores that fall in the following ranges:

    a) Between 70 and 80   b) More than 85

    c) Between 60 and 70   d) Less than 65

    e) Between 65 and 85   f)   More than 72

g) A score that deviates from the mean by more than 5

h) A score that deviates from the mean by more than 10

5. The heights of the 15,000 students of a certain university are normally distributed with a mean of 68 inches and a standard deviation of 4.7 inches. Determine the number of students that have the following heights:

a) Between 60 and 72 inches

b) Taller than 72 inches

c) Between 70 and 72 inches

d) Shorter than 72 inches

e) Shorter than 65 inches

f) Exactly 68 inches

g) A height that deviates from the mean by more than 3 inches

6. The heights of the 15,000 students of a certain university are normally distributed with a mean of 68 inches and a standard deviation of 4.7 inches. Determine the cutoffs for the following groups of heights:

a) Top 20%

b) Top 10%

c) Bottom 15%

d) Bottom 25%

e) Top 40%

f) Bottom 40%

7. The mean weight of a box of cereal filled by a machine is 15.0 ounces, with a standard deviation of 0.6 ounces. Assuming a normal weight distribution, find the percentage of the boxes that have the following weights:

a) Less than 14 ounces

b) Between 14 and 16 ounces

c) More than 16 ounces

d) Between 14.5 and 15.8 ounces

e) That deviate from 15.0 by more than 1 ounce

f) That deviate from 15.0 by more than 1.5 ounce

8. The amount of time required to assemble a certain product is normally distributed with a mean of 3.4 minutes and a standard deviation of 0.5 minutes. Find the probability that a randomly selected employee will require the following amount of time to assemble the product:

a) Less than 2 minutes

b) Between 3 and 4 minutes

c) Between 2 and 3 minutes

d) Less than 4 minutes

e) More than 4 minutes

f) More than 5 minutes

g) A time that deviates from 3.4 by more than 1 minute

h) A time that deviates from 3.4 by more than 2 minutes

9. The time it takes my favorite acrylic paint to dry is normally distributed with a mean of 2 hours and 45 minutes and a standard deviation of 30 minutes. Find the probability of the following drying times:

a) Between 2 and 3 hours   b) Less than 4 hours

c) Between 3 and 4 hours   d) More than 3 hours

e) Between $1\frac{1}{2}$ and $2\frac{1}{2}$ hours   f) More than 2 hours

g) A time that deviates from the mean by more than 30 minutes

h) A time that deviates from the mean by more than 40 minutes

10. The amount of time between taking a pain reliever and getting relief is normally distributed with a mean of 23 minutes and a standard deviation of 5 minutes. Find the probability that the time between taking the medication and getting relief is as follows:

a) Between 20 and 25 minutes   b) More than 20 minutes

c) Between 15 and 20 minutes   d) More than 23 minutes

e) Between 18 and 28 minutes   f) Less than 20 minutes

g) A time that deviates from the mean by more than 3 minutes

h) A time that deviates from the mean by more than 7 minutes

11. The mean weight of a box of cereal filled by a machine is 14.5 ounces, with a standard deviation of 0.5 ounces. Assuming a normal weight distribution, what percentage of the boxes will weigh the following amounts:

a) Less than 14 ounces   b) Between 14 and 16 ounces

c) More than 16 ounces   d) Between 14.5 and 15.8 ounces

e) A weight that deviates from the mean by more than 0.5 ounces

f) A weight that deviates from the mean by more than 0.8 ounces

12. The amount of time required to assemble a certain product is normally distributed with a mean of 3.6 minutes and a standard deviation of 0.6 minutes. Find the probability that a randomly selected employee will require the following amount of time to assemble the product:

a) Less than 2 minutes   b) Between 3 and 4 minutes

c) Between 2 and 3 minutes   d) Less than 4 minutes

e) More than 4 minutes   f) More than 5 minutes

13. The amount of time between taking a pain reliever and getting relief is normally distributed with a mean of 24 minutes and a standard deviation of 4 minutes. Find the probability that the time between taking the medication and getting relief is as follows:

a) Between 20 and 25 minutes   b) More than 20 minutes

c) Between 15 and 20 minutes   d) More than 23 minutes

e) Between 18 and 28 minutes   f) Less than 20 minutes

g) A time that deviates from the mean by more than 3 minutes

h) A time that deviates from the mean by more than 4.5 minutes

14. The amount of time required to assemble a certain product is normally distributed with a mean of 3.6 minutes and a standard deviation of 0.6 minutes. The company wishes to retrain its slower employees and reward the faster ones. Help the company find the cutoff assembly times that identify the following groups of employees:

a) Fastest 20%   b) Fastest 10%   c) Slowest 15%

d) Slowest 25%   e) Fastest 40%   f) Slowest 40%

15. A population is normally distributed with mean 27.4 and standard deviation 2.4.

a) Find the intervals representing one, two, and three standard deviations from the mean.

b) What percentage of the data lies in each of the intervals in a?

16. A population is normally distributed with mean 15.1 and standard deviation 3.1.

a) Find the intervals representing one, two, and three standard deviations from the mean.

b) What percentage of the data lies in each of the intervals in a?

2.5 Statistical Inference

Two uses for the normal curve have been described thus far: the approximation of binomial probabilities and the abbreviation of normally distributed data. There are many other situations in which this curve supplies essential information, and several of these, which are in very common use, are now described.

Recall the town discussed in Examples 1.3.3 through 1.3.6, 30% of whose population were fans of its pro basketball team. In that context, the question was asked of how likely a sample was to be misleading. Of course, the assumption that this proportion is known is unrealistic. Precise information of this sort is rarely available, and given the hypothetical nature of such numbers, the question of how misleading a sample can be is only of secondary interest. The reverse question is much more likely to be asked:

If 30 of 100 interviewees claimed to be fans, what inference can be made regarding the true proportion of fans in that town?

Problems of this type were first proposed and worked out by James Bernoulli (1654–1705) in his posthumously published book *Ars Conjectandi* (1713), and he suggested the format that the answer to such an a posteriori question should have, a format that is still very much in use today. Whereas it is impossible to determine the true proportion with absolute certainty on the basis of a sample, it is nevertheless possible to use the sample to trap this proportion within an interval so that one is *morally certain*, to use Bernoulli's phrase, that the proportion lies within the interval. Not only that, but Bernoulli showed how this moral certainty could be quantified in terms of a probability. Because he did not have the normal curve as a tool for estimating binomial probabilities, Bernoulli's mathematics turned out to be rather complex and impractical. Abraham De Moivre, who discovered the normal approximation to the binomial probabilities and incorporated it into the 1738 edition of his *Doctrine of Chances*, also showed how to express Bernoulli's intervals of moral certainty in terms of the normal curve. Today, the term *moral certainty* has been replaced by *confidence*, and Bernoulli's and De Moivre's intervals are called *confidence intervals*.

2.5.1   Confidence Interval for Proportion

For any small positive number α, the $(1 - \alpha)100\%$ *confidence interval* for the proportion p of a certain subpopulation, given by a sample of which the subpopulation accounts for proportion $\bar{p}$, has endpoints

$$\bar{p} \pm B\left(1 - \frac{\alpha}{2}\right)\sqrt{\frac{\bar{p}(1 - \bar{p})}{n}}$$

where $B(x)$ is obtained by reversing the reading of Table A. In other words,

$$B(0.95) = 1.645 \quad \text{because} \quad A(1.645) = 0.95$$
$$B(0.975) = 1.96 \quad \text{because} \quad A(1.96) = 0.975$$
$$B(0.99) = 2.326 \quad \text{because} \quad A(2.326) = 0.99$$
$$B(0.995) = 2.575 \quad \text{because} \quad A(2.575) = 0.995$$

Example 2.5.1. *Suppose a sample of 500 inhabitants contains 160 fans of the local basketball team. Construct a 95% confidence interval for the true proportion of fans in the town.*

Because the specified confidence level is $(1 - \alpha)100\% = 95\% = 0.95$, it follows that $\alpha = 0.05$ and so

$$B\left(1 - \frac{\alpha}{2}\right) = B\left(1 - \frac{0.05}{2}\right) = B(0.975) = 1.96$$

In addition, $\bar{p} = 160/500 = 0.32$ and $1 - \bar{p} = 1 - 0.32 = 0.68$. These numbers yield the endpoints

$$0.32 \pm 1.96 \cdot \sqrt{\frac{0.32 \cdot 0.68}{500}} \approx 0.32 \pm 0.04$$

of the confidence interval (0.28, 0.36). This means that we are 95% sure that the town's true proportion of fans lies between 28% and 36%. Depending on what stake one has in the issue under consideration, a confidence level of 95% may be insufficient. To obtain a 99% level of confidence, it is necessary to set $\alpha = 1\% = 0.01$, so that

$$B\left(1 - \frac{\alpha}{2}\right) = B\left(1 - \frac{0.01}{2}\right) = B(0.995) = 2.575$$

This yields the endpoints

$$0.32 \pm 2.575 \cdot \sqrt{\frac{0.32 \cdot 0.68}{500}} \approx 0.32 \pm 0.05$$

with a resulting confidence interval of (0.27, 0.37), which is unfortunately wider than the 95% interval of (0.28, 0.34) obtained above. Thus, the additional confidence comes at the cost of less accuracy. More accuracy can be obtained if one is willing to relinquish some of the certainty. For example, to obtain the 90% confidence interval, we set $\alpha = 10\% = 0.1$, so that

$$B\left(1 - \frac{\alpha}{2}\right) = B\left(1 - \frac{0.1}{2}\right) = B(0.95) = 1.645$$

This yields the endpoints

$$0.32 \pm 1.645 \cdot \sqrt{\frac{0.32 \cdot 0.68}{500}} \approx 0.32 \pm 0.03$$

with a resulting confidence interval of (0.29, 0.35). Table 2.5 lists the above three intervals and clarifies the inverse relation between the confidence level and the interval's width. The decision of which confidence level should be used must remain subjective. The 95% confidence level is the default level and also the one most commonly used by experimenters.

Confidence level	Interval
99%	(0.27, 0.37)
95%	(0.28, 0.36)
90%	(0.29, 0.35)

Table 2.5   Some standard confidence levels

All the above intervals may seem unsatisfactory because there is a considerable gap between the two endpoints. The difference between 29% and 35% is substantial, as is that between 27% and 37%. The term

$$B\left(1 - \frac{\alpha}{2}\right)\sqrt{\frac{\bar{p}(1 - \bar{p})}{n}} \tag{2.5}$$

is of course responsible for the width of the confidence interval. The variables α and n can be controlled by the pollster, and their effect on the width is discussed momentarily. The term $\bar{p}$, on the other hand, is unpredictable and so it is eliminated by reverting to the worst case scenario, which happens to occur when $\bar{p} = 0.5$. In other words, because

$$B\left(1 - \frac{\alpha}{2}\right)\sqrt{\frac{\bar{p}(1 - \bar{p})}{n}} \leq B\left(1 - \frac{\alpha}{2}\right)\sqrt{\frac{0.5(1 - 0.5)}{n}} \leq \frac{B\left(1 - \frac{\alpha}{2}\right)}{2\sqrt{n}}$$

we define

$$\text{MOE} = \frac{B\left(1 - \frac{\alpha}{2}\right)}{2\sqrt{n}}$$

as the *margin of error* (MOE) of the sample, knowing that if we succeed in keeping it small, then the actual but unpredictable margin of (2.5) will also be small. In the previous example, in the case where $\alpha = 5\%$, the margin of error is

$$\frac{B(0.975)}{2\sqrt{500}} = \frac{1.96}{2\sqrt{500}} \approx 0.044$$

which is quite close to the actual error bound of 0.04.

In general, the margin of error should be thought of as the measure of accuracy of a statistically based estimation of some true proportion. The smaller the margin of error, the more accurate the test. It should be of interest to see how variations in the level of confidence affect the margin of error. The table below indicates the comparative widths of the confidence intervals associated with various standard confidence levels but identical sample sizes:

Confidence level:	99%	98%	95%	90%
α:	1%	2%	5%	10%
Width of confidence interval:	$\dfrac{2.58}{2\sqrt{n}}$	$\dfrac{2.33}{2\sqrt{n}}$	$\dfrac{1.96}{2\sqrt{n}}$	$\dfrac{1.65}{2\sqrt{n}}$
Comparison of width	132%	119%	100%	84%

Thus, to increase the confidence level of the above test from 95% to 99% (without changing the sample), it is necessary to increase the interval width by 32%. Unfortunately, there are no guidelines for answering the trade-off question of whether the increase in confidence is justified by the accompanying decrease in the accuracy of the estimation. These decisions too must remain subjective.

It is also possible to narrow the width of the confidence interval by increasing the sample size. Because of the square root in the denominator of the margin of error, doubling the accuracy (or halving the width) of the estimating interval calls for quadrupling the sample size, which can be quite expensive. In the $\alpha = 5\%$ case of the discussion above, where a sample of 500 resulted in an interval of width 8%, it would require sample sizes of 1000, 2000, and 4000 to obtain interval widths of 4%, 2%, and 1%, respectively.

It should be useful to have a method for computing the sample size that yields a prespecified accuracy. This is given by the formula

$$n = \left(\frac{B(1 - \alpha/2)}{w} \right)^2$$

where w is the desired width of the confidence interval. Accordingly, if a confidence level of 98% and a width of 4% are desired, then the required sample size is

$$\left(\frac{B(0.99)}{0.04} \right)^2 = \left(\frac{2.326}{0.04} \right)^2 \approx 3382$$

Once again, there are no general guidelines for resolving the trade-off between the confidence level and the cost of the statistical test.

2.5.2   Confidence Interval for Mean

Confidence intervals can also be used to estimate unknown averages of large population. This is the sort of problem that arises in, say, estimating the average annual income of a large group of people on the basis of a sample. Here, too, no sample can be guaranteed to yield the true mean of a large population, and it is necessary to make do with confidence intervals. Suppose a sample of n (in practice, n must be at least 30) data from a large population with unknown average m has average $\bar{x}$ and standard deviation s. Then the numbers

$$\bar{x} \pm B\left(1 - \frac{\alpha}{2} \right) \frac{s}{\sqrt{n}}$$

are the endpoints of a $(1 - \alpha)100\%$ confidence interval for the population average m. For example, if a random sample of 300 families in a certain town report an average income of $34,355 with a standard deviation of $9811, then the numbers

$$34,355 \pm B\left(1 - \frac{0.05}{2} \right) \cdot \frac{9811}{\sqrt{300}} = 34,355 \pm 1.96 \cdot \frac{9811}{\sqrt{300}} \approx 34,355 \pm 1110$$

constitute a 95% confidence interval for the average family income over the entire population. In other words, we are 95% confident that the average family income for the town lies between \$33,245 and \$35,465. The author, for one, considers this to be a much more satisfactory answer than the 95% confidence interval of (28%, 36%) for the proportion of a team's fans in a different town. The difference between 28% and 36% seems much larger than the difference between \$33,245 and \$35,465. Perhaps this is because 35,465 is only 7% greater than 33,245, whereas 36 is 29% greater than 28.

2.5.3   Hypothesis Testing for Mean

Another use of the normal distribution is that of hypothesis testing. Consider, for example, the case of a manufacturer's cans of peas in which the net weights are 16 ounces. Such a company is likely to check on the net content of its product. In doing so, attention must be paid to two different potential problems. The cans could in reality contain less than 16 ounces, in which case the manufacturer could face charges of false advertising. On the other hand, the cans could contain more than 16 ounces, in which case the company is placing itself at a disadvantage relative to its competitors. Thus, the company has an interest in making sure that the net contents of its can are as close to the advertised 16 ounces as possible. Because of the intrinsic randomness inherent in our universe in general and in all manufacturing processes in particular, no two cans can have exactly the same weight and the tester must be satisfied with verifying that the *average* contents are as stated rather than the contents of each can. Thus, the tester uses samples to test the so-called *(null) hypothesis:*

H_0: *The mean weight of a can is 16 ounces*

The idea that underlies hypothesis testing is quite straightforward. A sample of the product in question is examined. If there is a considerable difference between the average weight of the sampled cans and 16 ounces, then the sample is considered as strong evidence that the true average weight is *not* 16 ounces. Of course, it is necessary to quantify the notion of *considerable difference*, and this is where the normal distribution comes in. Given a population with standard deviation s and any positive number d, the quantity

$$2\left(1 - A\left(\frac{d}{s/\sqrt{n}}\right)\right)$$

turns out to be the probability that an n-sample will have a mean that differs from the population's *mean* of m by at least as much as d. Consequently, we define

$$p\text{-value} = 2\left(1 - A\left(\frac{|\bar{x} - m|}{s/\sqrt{n}}\right)\right)$$

Suppose that in the above context 100 cans of peas are examined and their net weights average 15.88 ounces with a standard deviation of 0.42 ounces. Then,

$$p\text{-value} = 2\left(1 - A\left(\frac{|15.88 - 16|}{0.42/10}\right)\right) = 2(1 - A(2.86))$$

$$= 2(1 - 0.9979) = 0.0042 \approx 0.4\%$$

To understand the significance of the p-value, note that the low sample mean of 15.87 ounces could be the result of two causes:

1. The average weight of the boxes *is* 16 ounces and the sample is a fluke.
2. The average weight of the boxes really is less than 16 ounces.

The p-value is essentially the probability of the first of these two alternatives. As such, it measures the test's *significance*. The smaller the p-value of a statistical test, the less likely it is to be a fluke and hence the more significant, or meaningful, is the result of the test. The above obtained 0.4% means that if we conclude on the basis of this sample that the average weight really is different from 16 ounces, then there is only a chance of 0.4% of making a mistake. This fraction of a percent is so small that most people would be willing to accept that kind of a risk. In practice, most experimenters will accept a p-value (i.e., risk of drawing the wrong conclusion) that is less than $0.05 = 5\%$ as a sufficiently strong grounds for rejecting a null hypothesis.

Example 2.5.2. *Whole milk is supposed to contain 3% fat. To test whether this is true for the milk sold by the How Now Brown Cow dairy, 30 of their containers were opened and it was found that their fat contents average 3.1% with a standard deviation of 0.3%. Find the p-value of this test and decide whether this test indicates the fat content is not 3%.*

The p-value of this test is

$$2\left(1 - A\left(\frac{3.1 - 3}{0.3/\sqrt{30}}\right)\right) = 2(1 - A(1.83)) = 2(1 - 0.9664) \approx 6.7\%$$

which exceeds 5%. Hence, the sample cannot be understood as contradicting the null hypothesis that the fat content of the whole milk sold by the How Now Brown Cow dairy is 3%.

Example 2.5.3. *Suppose the sample of Example 2.5.2 had an average fat content of 3.12% instead of 3.1%. How would that affect the conclusion to be drawn from the test?*

The p-value of the test would be

$$2\left(1 - A\left(\frac{3.12 - 3}{0.3/\sqrt{30}}\right)\right) = 2(1 - A(2.19)) = 2(1 - 0.9857) \approx 2.9\%$$

which is less than 5%. Under these circumstances it would be proper to conclude that the average fat content of the dairy's whole milk is not 3%. This makes sense because, all other things being equal, the greater the difference between the hypothesized average and the sample average, the more likely the hypothesized average is to be wrong.

Example 2.5.4. *Suppose the sample of Example 2.5.3 had a standard deviation of 0.5% instead of 0.3%. How would that affect the conclusion to be drawn from the test?*

The p-value of the test would be

$$2\left(1 - A\left(\frac{3.1 - 3}{0.5 / \sqrt{30}}\right)\right) = 2(1 - A(1.10)) = 2(1 - 0.8643) \approx 27.1\%$$

which is more than 5%. Consequently, the test does not cast doubt on the claim that the fat content is indeed 3%. Notice that in the transition from the previous example to this one, the p-value, which is the likelihood of this test to be a fluke, has increased. Because the evidence of the original example was deemed too weak to reject the 3% hypothesis, so must be every test with an even greater p-value, all else being equal.

Example 2.5.5. *Suppose Example 2.5.2 had had a sample of size 40 instead of 30. How would that have affected the conclusion to be drawn from the test?*

The p-value would have been

$$2\left(1 - A\left(\frac{3.1 - 3}{0.3 / \sqrt{40}}\right)\right) = 2(1 - A(2.11)) = 2(1 - 0.9826) \approx 3.5\%$$

which is less than 5%. Under these circumstances it would have been proper to conclude that the average fat content of the dairy's whole milk is not 3%. This illustrates the beneficial effect of increasing the sample size. The same average and standard deviation become more discriminating when they come from a larger sample.

2.5.4   Hypothesis Testing for Proportion

The process of hypothesis testing can be applied to the proportion of subpopulations as well. For example, it might be desired to *test* the widely held belief that the proportion of fans in the aforementioned town (see Section 1.3) is indeed $0.3 = 30\%$. To do this, one might interview a random sample of inhabitants and compare the proportion of fans in the sample to the putative proportion of 0.3. Once again, there is a p-value to help gauge strength of the evidence. Specifically, to test the null hypothesis

H_0: *The proportion of a certain subpopulation is p*

one takes a random sample (of size n) and computes the proportion $\bar{p}$ of the subpopulation present in it. In this context, the likelihood of an error if the difference between p and $\bar{p}$ is used to discredit the null hypothesis is given by

$$p\text{-value} = 2\left(1 - A\left(\frac{|\bar{p} - p|}{\sqrt{p(1-p)/n}}\right)\right)$$

Example 2.5.6. *Two hundred inhabitants of the aforementioned town are randomly selected and interviewed. Seventy of them claim to be fans. Is this strong enough evidence to conclude that the widely held belief that 30% of the townspeople are fans is wrong?*

Here $\bar{p} = 70/200 = 0.35$ so that $1 - \bar{p} = 0.65$. Consequently,

$$p\text{-value} = 2\left(1 - A\left(\frac{0.35 - 0.30}{\sqrt{0.35 \cdot 0.65 / 200}}\right)\right) = 2(1 - A(1.48)) = 2(1 - 0.9306) \approx 13.8\%$$

Because 13.8% exceeds the 5% cutoff level mentioned above, most statisticians would consider it a rash act to reject the 30% hypothesis on the basis of this evidence alone.

Example 2.5.7. *Suppose in the previous example the sample size was 400 rather than 200. How would that affect the conclusion to be drawn?*

In this case the p-value is

$$2\left(1 - A\left(\frac{0.35 - 0.30}{\sqrt{0.35 \cdot 0.65 / 400}}\right)\right) = 2(1 - A(2.10)) = 2(1 - 0.9821) = 0.0358 \approx 3.6\%$$

Because 3.6% is less than 5%, most statisticians would accept the result of this poll as indicating that the true proportion of fans is no longer 30%.

EXERCISES 2.5

1. Of a random sample of 120 shoppers in a certain town, 86 used coupons. Construct a $(1 - \alpha)100\%$ confidence interval for the true proportion of coupon users where α is (a) 1%, (b) 5%, and (c) 10%. At each of these levels of confidence, determine the margin of error as well as the smallest sample size that guarantees an interval of width 0.04.

2. An examination of 250 randomly selected police files found that 156 crimes involved firearms. Construct a $(1 - \alpha)\%$ confidence interval for the true proportion of crimes that involve firearms where α is (a) 1%, (b) 5%, and (c)

10%. At each of these levels of confidence, determine the margin of error as well as the smallest sample size that guarantees an interval of width 0.03.

3. A certain medical screening test was found to misdiagnose 37 of 300 randomly selected test cases. Construct a $(1 - \alpha)100\%$ confidence interval for the true proportion of test failures where α is (a) 1%, (b) 5%, and (c) 10%. At each of these levels of confidence, determine the margin of error as well as the smallest sample size that guarantees an interval of width 0.02.

4. A new drug is tested on 400 randomly selected patients and is found to be effective in 343 cases. Construct a $(1 - \alpha)100\%$ confidence interval for the true effectiveness of the drug where α is (a) 1%, (b) 5%, and (c) 10%. At each of these levels of confidence, determine the margin of error as well as the smallest sample size that guarantees an interval of width 0.05.

5. A survey of 150 randomly selected homeowners in a certain town finds that 10 of them do not have home insurance. Construct a $(1 - \alpha)100\%$ confidence interval for the true proportion of uninsured homeowners in that town where α is (a) 1%, (b) 5%, and (c) 10%. At each of these levels of confidence, determine the margin of error as well as the smallest sample size that guarantees an interval of width 0.02.

6. Twenty-five boxes of a certain kind of cereal are opened and it is determined that their average net weight is 14.8 ounces with a standard deviation of 0.15 ounces. Assuming that these weights are normally distributed, form a $(1 - \alpha)100\%$ confidence interval for the true mean weight where α is (a) 1%, (b) 5%, and (c) 10%.

7. A study of 45 randomly selected workers finds that the assembly of a certain product requires, on the average, 3.6 minutes, with a standard deviation of 1.5 minutes. Construct a $(1 - \alpha)100\%$ confidence interval for the true mean assembly time where α is (a) 1%, (b) 5%, and (c) 10%.

8. Seventy-five cereal boxes of a certain brand are found to average a net weight of 15.7 ounces with a standard deviation of 0.08 ounces. Construct a $(1 - \alpha)100\%$ confidence interval for the true average net weight where α is (a) 1%, (b) 5%, and (c) 10%.

9. A survey of 250 randomly selected moviegoers has found that, on the average, each spent $242 on movies during the past year, with a standard deviation of $21. Construct a $(1 - \alpha)100\%$ confidence interval for the true mean annual expenditure of the typical moviegoer where α is (a) 1%, (b) 5%, and (c) 10%.

10. The average salary of a random sample of 200 employees of a certain large company is $29,342 with a standard deviation of $2531. Construct a $(1 - \alpha)100\%$ confidence interval for the true average salary where α is (a) 1%, (b) 5%, and (c) 10%.

11. To test the belief that 60% of shoppers in a certain town use coupons, 120 randomly selected shoppers were observed and 86 of those were found to use coupons. What is the p-value of this test? What conclusion should be drawn at the level of significance of (a) $\alpha = 0.5\%$, (b) $\alpha = 1\%$, and (c) $\alpha = 10\%$?

12. In previous years the percentage of crimes that involves firearms has held steady at 50%. An examination of 250 recent police files found that 156 of them involved firearms. Does this indicate that the percentage in question changed? Draw a separate conclusion at each of the significance levels of (a) $\alpha = 0.5\%$, (b) $\alpha = 1\%$, and (c) $\alpha = 10\%$.

13. A certain medical screening test is known to be 90% effective. A new procedure is found to be correct on 190 of 200 people. Does this indicate that the new procedure has a different effectiveness rate? Draw a separate conclusion at each of the significance levels of (a) $\alpha = 0.5\%$, (b) $\alpha = 1\%$, and (c) $\alpha = 10\%$.

14. A survey found that of 185 voters who came to a certain booth, 99 were women. Does this indicate that the proportion of females among voters is different from 50%? Draw a separate conclusion at each of the significance levels of (a) $\alpha = 0.5\%$, (b) $\alpha = 1\%$, and (c) $\alpha = 10\%$.

15. For years the proportion of coffee drinkers among American adults has been 75%. A study finds that of 300 randomly selected adult Americans, 212 drink coffee. Has a change occurred? Draw a separate conclusion at each of the significance levels of (a) $\alpha = 0.5\%$, (b) $\alpha = 1\%$, and (c) $\alpha - 10\%$.

16. The assembly of a certain product requires, on the average, 3.6 minutes. A new method is tried out on 50 employees and is found to require, on the average, 3.1 minutes with a standard deviation of 1.5 minutes. Is there enough evidence to indicate that the new method is substantially different? Draw a separate conclusion at each of the significance levels of (a) $\alpha = 0.5\%$, (b) $\alpha = 1\%$, and (c) $\alpha = 10\%$.

17. I am a painter and the amount of time it takes my favorite acrylic paint to dry on a fair day is, on the average, 2.5 hours. I test a random sample of 30 cans of a different kind of paint and find that the average drying time is 2 hours and 10 minutes with a standard deviation of 45 minutes. Does the evidence indicate that the drying times are different? Draw a separate conclusion at each of the significance levels of (a) $\alpha = 0.5\%$, (b) $\alpha = 1\%$, and (c) $\alpha = 10\%$.

18. A company claims that its cereal boxes contain, on the average, 16 ounces of cereal. Twenty-five boxes are examined by that company's test product division and found to contain, on the average, 15.9 ounces with a standard deviation of 0.2 ounces. Does the evidence contradict the claim? Draw a separate conclusion at each of the significance levels of (a) $\alpha = 0.5\%$, (b) $\alpha = 1\%$, and (c) $\alpha = 10\%$.

19. To determine whether the average expenditure per person on movies has changed, a random sample of 30 people was asked to compare their expenditures in 1997 year to those of 1996. It was found that the average expenditure increased by $20 with a standard deviation of $55 in the reported differences. Does the sample indicate a change in people's spending habits? Draw a separate conclusion at each of the significance levels of (a) $\alpha = 0.5\%$, (b) $\alpha = 1\%$, and (c) $\alpha = 10\%$.

20. A manufacturer of small electric motors asserts that on the average they draw 1.8 amperes under normal load conditions. A sample of 25 of the motors was tested and it was found that the mean current was 1.96 amperes with a standard deviation of .32 amperes. Does the evidence bear out the manufacturer's claim. Draw a separate conclusion at each of the significance levels of (a) $\alpha = 0.5\%$, (b) $\alpha = 1\%$, and (c) $\alpha = 10\%$.

2.6   The Central Limit Theorem

One of the reasons that the normal curve is such a popular tool in statistical inference is that even when a population is not normally distributed, its sample averages do tend to be so distributed. This fact, which is known as the Central Limit Theorem and was discovered in 1810 by P.-S. Laplace (1749–1827), implies the near universality of the testing methods described in the previous section. It is now illustrated with several examples.

The data {1, 2, 3, 4, 5, 6} are not normally distributed—there is neither a central aggregation nor a tapering off of the data toward the ends. Visually, the histogram of these data is flat (Fig. 2.19). Statistical inference, however, is concerned with samples and their averages. The simplest possible nontrivial sampling that can be done calls for selecting two data at random. There are 15 such potential samples, and they and their sample averages are listed below:

Two-sample:	{1, 2}	{1, 3}	{1, 4}	{1, 5}	{1, 6}	{2, 3}	{2, 4}	{2, 5}	{2, 6}
Sample average:	1.5	2	2.5	3	3.5	2.5	3	3.5	4

Two-sample:	{3, 4}	{3, 5}	{3, 6}	{4, 5}	{4, 6}	{5, 6}
Sample average:	3.5	4	4.5	4.5	5	5.5

These sample averages are regarded as data themselves and as such have a histogram that appears in Figure 2.20. Note that whereas this graph is still not bell shaped, it is closer to being so than the completely flat histogram of Figure 2.19. This is one of the implications of the central limit theorem. Regardless of the shape (of the histogram) of the original data, histograms of sample averages tend to assume the bell shape. This assertion is also demonstrated by Figure 2.21, which tabulates the two-sample averages of the data {1, 2, 3, 4, 5, 6, 7, 8}. The fit in both of these examples is some-

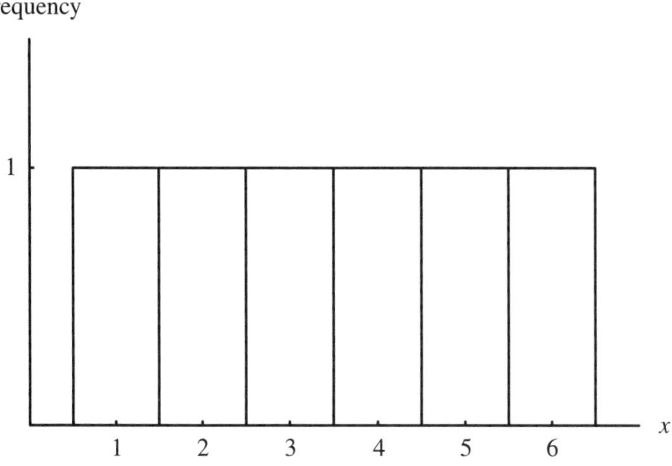

Figure 2.19   *A flat histogram.*

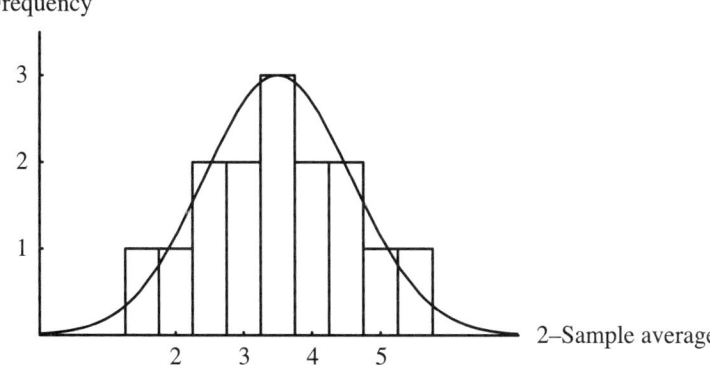

Figure 2.20   *n = 6. The histogram of the 2-sample averages of the data {1, 2, 3, 4, 5, 6}.*

what forced. The histograms of the two-sample averages are actually triangular, a shape that is still closer to the ideal bell shape than to the rectangle we started out with in Figure 2.20. However, the characteristic tapering off of the bell's rim becomes evident when averages of three-samples are examined. Figure 2.22 displays the histogram of the three-sample averages of the data {1, 2, 3, 4, 5, 6, 7, 8, 9}; these three samples and their averages are listed below:

Three-sample: {1, 2, 3}   {1, 2, 4} $\cdots$ {7, 8, 9}
Sample average: 2 $7/3$ $\cdots$ 8

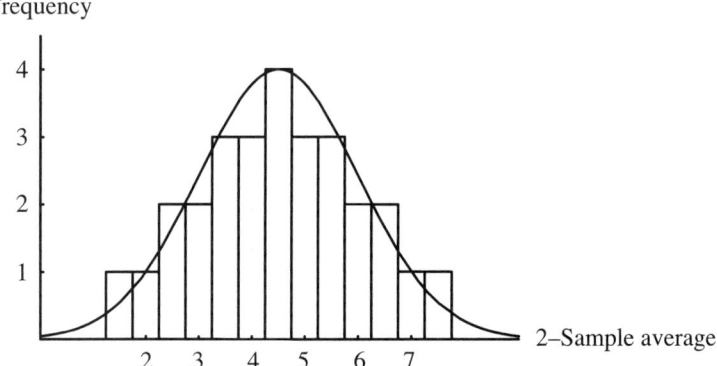

Figure 2.21   n = 8. *The histogram of the 2-sample averages of the data {1, 2, 3, 4, 5, 6, 7, 8}.*

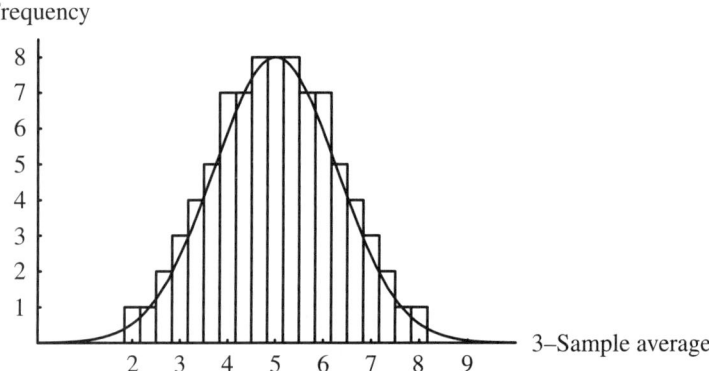

Figure 2.22   n = 9. *The histogram of the 3-sample averages of the data {1, 2, 3, 4, 5, 6, 7, 8, 9}.*

This trend toward normality is also evident in Figure 2.23, which displays the histogram of the four-sample averages of the data {1, 2, 3, 4, 5, 6, 7, 8, 9, 10}. It is impossible to give an accurate statement of the Central Limit Theorem in this context because it is too technical. Informally, this theorem, which is the foundation of much of statistical practice, says that

For any population with average m and standard deviation s and for any integer $n \geq 30$, the averages of the n-samples are (approximately) normally distributed with average m and standard deviation $s / \sqrt{n}$).

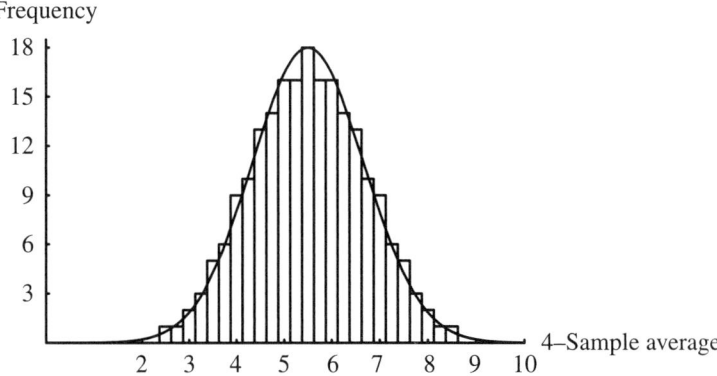

Figure 2.23   n = 10. *The histogram of the 4-sample averages of the data* {1, 2, 3, 4, 5, 6, 7, 8, 9, 10}.

EXERCISES 2.6

For each of the populations in Exercises 1–13,

a) Construct the frequency table of the population and draw its histogram.

b) Compute the average and the standard deviation of the population.

c) Construct a frequency table for the two-sample averages and draw its histogram.

d) Compute the average and standard deviation of the two-sample averages.

e) Construct a frequency table for the three-sample averages and draw its histogram.

f) Compute the average and standard deviation of the three-sample averages.

1. {1, 2, 3, 4}	2. {2, 3, 4, 5}	3. {1, 2, 5, 6}
4. {1, 2, 2, 3}	5. {1, 2, 3, 3}	6. {1, 2, 2, 2}
7. {2, 2, 2, 2}	8. {1, 2, 3, 4, 5}	9. {1, 2, 2, 3, 4}
10. {1, 2, 2, 3, 3}	11. {1, 2, 2, 2, 3}	12. {1, 2, 2, 2, 2}
13. {3, 3, 3, 3, 3}		

2.7   Activities/Websites

1. Collect and tabulate the heights (in inches) of the students in your class. Compute the mean and the standard deviation. Use Figure 2.12 to check whether these data are normally distributed.

2. Collect and tabulate the weights (in pounds) of the students in your class. Compute the mean and the standard deviation. Use Figure 2.12 to check whether these data are normally distributed.

3. Let each student in class flip a coin and keep flipping, if necessary, until the first *heads* is obtained. Record the number of flips done by each student and have the students repeat the experiment until at least 30 data are obtained. Compute the average and the standard deviation of the data. (Theoretically, they should be 2 and $\sqrt{2}$, respectively.) Repeat the experiment with three times as many data.

4. Let each student in class toss a die and keep tossing, if necessary, until the first six is obtained. Record the number of tosses done by each student and have the students repeat the experiment until at least 30 data are obtained. Compute the average and the standard deviation of the data. (Theoretically, they should be 6 and $\sqrt{30}$, respectively.) Repeat the experiment with three times as many data.

5. It would seem that if a coin is spun, then the probability of its coming up *heads* is 50%. On the other hand, it could be argued that because the faces are different, perhaps the 50% is incorrect. Test the 50% hypothesis at $\alpha = 5\%$ by spinning the same coin 100 times.

Chapter 3
Voting Systems

3.1 Introduction

We rank everything. We rank movies, bands, songs, racing cars, politicians, professors, sports teams, the plays of the day. We want to know "who's number one?" and who isn't. As a local sports writer put it, "Rankings don't mean anything. Coaches continually stress that fact. They're right, of course, but nobody listens. You know why, don't you? People love polls. They absolutely love 'em" (Woodling, 2004). Some rankings are just for fun, but people at the top sometimes stand to make a lot of money. A study of films released in the late 1970s and 1980s found that if a film is one of the five finalists for the Best Picture Oscar at the Academy Awards, the publicity generates (on average) about $5.5 million in additional box office revenue. Winners make, on average, $14.7 million in additional revenue (Nelson et al., 2001). In today's inflated dollars, the figures would no doubt be higher. Actors and directors who are nominated (and who win) should expect to reap rewards as well, because the producers of new films are eager to hire Oscar winners.

This chapter is about the procedures that are used to decide who wins and who loses. Developing a ranking can be a tricky business. Political scientists have, for centuries, wrestled with the problem of collecting votes and molding an overall ranking. To the surprise of undergraduate students in both mathematics and political science, mathematical concepts are at the forefront in the political analysis of voting procedures. Mathematical tools are important in two ways. First, mathematical principles are put to use in the scoring process that ranks the alternatives. Second, mathematical principles are used to describe the desired properties of a voting procedure and to measure the strengths and weaknesses of the procedures. Mathematical concepts allow us to translate important, but vague, ideas like "logical" and "fair" into sharp formally defined expressions that can be applied to voting procedures.

In this chapter, we consider a number of voting procedures and we discuss their strengths and weaknesses.

3.2   Mathematical Concepts

3.2.1   Transitivity

Real numbers are transitive. Every school child knows that transitivity means that

$$\text{If } x > y \text{ and } y > z, \text{ then } x > z \tag{3.1}$$

The symbol > means "greater than" and ≥ means "greater than or equal to." Both of these are transitive binary relations. "Binary" means only two numbers are compared against each other, and transitivity means that many numbers can be linked together by the binary chain. For example, we can work our way through the alphabet. If $a > b$, and $b > c$, and ... $x > y$, then $a > z$.

If a, b, and c represent policy proposals or football teams being evaluated, we expect that voters are able to make binary comparisons, declaring that one alternative is preferred to another or that they are equally appealing. The parallel between "greater than" and "preferred to" is so strong, in fact, that we use the symbol $\succ$ to represent "preferred to." The subscript i indicates that we are talking about a particular voter, so $x \succ_i y$ means that voter i prefers x to y and $x \succeq_i y$ means that x is as good as y. If i is indifferent, we write $x \approx_i y$.

Using the binary relations $\succ_i$ and $\succeq_i$, it is now possible to write down the idea of transitive preferences. If a person's preferences are transitive, the following holds:

$$\text{If } x \succ_i y \text{ and } y \succ_i z, \text{ then } x \succ_i z \tag{3.2}$$

This can be written more concisely as

$$x \succ_i y \succ_i z \tag{3.3}$$

According to William Riker, a pioneer in the modern mathematical theory of voting, transitivity is a basic aspect of reasonable human behavior (Riker, 1982). Someone who prefers green beans to carrots and also prefers carrots to squash is expected to prefer green beans to squash. One of the truly surprising—even paradoxical—problems that we explore in this chapter is that social preferences, as expressed through voting procedures, may not be transitive. The procedure of majority rule, represented by the letter M, can compare alternatives so that: $x \succ_M y \succ_M z$, and yet x does not defeat z in a majority election. The fact that individuals may be transitive but social decision procedures are not has been a driving force in voting research.

3.2.2   Ordinal Versus Cardinal Preferences

If $x \succ_i y$, we know that x is preferred to y, but we don't know by "how much." These are called ordinal preferences because they contain information only about ordering and not magnitude. Can the magnitude of the difference be measured? The proponents of an alternative model, called cardinal preferences, have developed ingenious mathematical techniques to measure preferences.

The proponents of ordinal preferences argue that all the information about the tastes of the voters that is worth using is encapsulated in a statement like $x \succ_i y$. Even if we could measure the gap in the desirability between x and y in the eyes of voter i, it would still not solve the problem that the gaps observed by voters i and j would not be comparable. There is no rigorous mathematically valid way to express the idea that i likes policy x "twice as much" as voter j, even though such an interpersonal comparison is tempting. As a result, the ordinalists argue that no voting procedure should be designed with the intention of trying to measure "how much" more attractive x is than y. Only the ranking should be taken into account.

The approaches based on cardinal preferences offer the possibility of very fine-grained and, at least on the surface, precise social comparisons. If we had a reliable measure of preferences on a cardinal scale, perhaps it would be possible to say that x is 0.3 units more appealing than y. The most widely used method of deriving these preferences is the Von Neumann-Morgenstern approach, which is used in game theory (see Chapter 4). The theory that justifies the measurement techniques is somewhat abstract and complicated. Even in a laboratory setting, these preferences have proven difficult to measure reliably.

For whatever reasons, almost all voting procedures in the world today use ordinal methods. We ask voters to declare their favorite or to rank the alternatives first, second, and third. As we discuss the weaknesses of various voting methods, readers may be tempted to wonder if the use of cardinal preference information might lead to an easy fix. We believe that such quick fixes are unlikely to be workable, but they are often very interesting. The fact that one cannot compare preference scores across individual voters should never be forgotten, because many proposals implicitly assume that such comparisons are meaningful.

EXERCISES 3.2

1. Consider three lunch items, *steak, fish*, and *eggplant*. If Joe prefers steak to fish, we write *steak* $\succ_{Joe}$ *fish*. Joe also prefers *fish* to *eggplant*. If Joe's preferences are transitive, should we conclude *steak* $\succ_{Joe}$ *eggplant* or *eggplant* $\succ_{Joe}$ *steak*?

2. Suppose Jennifer tells us *steak* $\succ_{Jennifer}$ *fish* and *steak* $\succ_{Jennifer}$ *eggplant*. Can we say whether *steak* $\succ_{Jennifer}$ *fish* or *fish* $\succ_{Jennifer}$ *steak*?

3. If *pink* $\succ_{Mary}$ *red* and *red* $\succ_{Mary}$ *green*, and then Mary says she prefers *green* to *pink*, would her choices seem reasonable to you? Why?

3.3   The Plurality Problem

By far the most commonly used election procedure in the United States is simple plurality rule. Each eligible voter casts one vote, and the winner is the candidate who receives the most votes. In terms borrowed from horse racing, it is sometimes called a "first past the post" procedure. The winner need only have more support than the second-placed competitor. This method is used in elections in other countries (e.g., Great Britain), and it is the method of balloting used in the selection of winners of the prestigious Oscar Awards, which are offered by the Academy of Motion Picture Arts and Sciences.

3.3.1   Two Candidates? No Problem!

If there are only two candidates, then the plurality rule is a good method of choice. It is, in fact, equivalent to majority rule. Suppose the alternatives are x and y and the voters are $N = \{1, 2, 3, ..., n\}$. Voters for whom $x \succ_i y$ will vote for x. In *majority rule*, which we label M, an alternative with the support of more than one half of the voters is the winner. Using the notation

$$| \{i \in N : condition\} |$$

to indicate "number of elements in N for which *condition* is true," majority rule is stated as

$$\frac{| \{i \in N : x \succ_i y\} |}{n} > \frac{1}{2} \text{ implies } x \succ_M y \tag{3.4}$$

The voters who oppose x, the ones for whom $y \succ_i x$, or are indifferent, $x \approx_i y$, are viewed as the opposition.

In contrast, under *plurality rule*, which we label P, the candidate preferred by a greater number of voters wins, even if that candidate has less than one half. Plurality rule is formally defined as

$$| \{i \in N : x \succ_i y\} | > | \{i \in N : y \succ_i x\} | \text{ implies } x \succ_P y \tag{3.5}$$

Note this ignores indifferent voters. In contrast, in a majority system the indifference is treated as opposition. If indifferent voters exist but they abstain from voting, then plurality and majority again coincide. Only if indifferent voters are given a way to register their indifference, such as casting a "tie" vote, do the two methods diverge.

There is a mathematical proof of the superiority of majority rule known as May's theorem (May, 1952). Many people have been confused by the fact that May is actually discussing the system that we have defined as plurality rule. In the system that May calls majority rule, votes in favor of two alternatives are collected and the one with the most favorable votes wins (the indifferent voters are not counted as opposition). May's theorem states that when there are two alternatives, majority (what we call plurality) rule is the only decisive procedure that is consistent with these three elementary properties:

1. **Anonymity:** Each person's vote is given the same weight.

2. **Neutrality:** Relabeling the alternatives (switching the titles x and y) in voter preferences causes an equivalent relabeling of the outcome (the alternatives

are undifferentiated by their labels, sometimes called *undifferentiatedness* of alternatives).

3. **Monotonicity:** The electoral system is positively responsive in the following two senses:

   a. **strong monotonicity:** if there is a tie and then one voter elevates x in his preferences (that is, changes from $y \succ_i x$ to $x \approx_i y$ or $x \succ_i y$ or from $x \approx_i y$ to $x \succ_y y$), then x must be the winner;

   b. **weak monotonicity:** if x is the winner and one voter elevates x in his preferences, then x must remain the winner.

The formalization and proof of May's theorem is discussed in the exercises.

May's theorem is one of the most encouraging results in the study of voting procedures. It points our attention at a particular method of decision making, and it formally spells out the virtuous properties of the procedure.

3.3.2 Shortcomings of Plurality Rule with More Than Two Candidates

Because majority rule has so much appeal with two candidates, there is a natural tendency to try to stretch it to apply in an election with more candidates. The two most important methods through which this has been tried are *pure plurality rule* and the *majority/runoff election system*. In pure plurality, the winner is the one with the most votes, even if the vote share is only 5% or 10%. In the majority/runoff system, all the candidates run against each other and the winner is required to earn more than one half of all votes cast. If no candidate wins a majority, then the top two vote getters are paired off against each other in a runoff election. This system does not typically include "none of the above" as an alternative in either stage, and so there is no method for voters to register indifference (and hence, majority and plurality rule with two candidates imply the same results).

The two-stage majority/runoff system is used in some American state and local elections and in French national elections. This system has obvious flaws. The candidates who have the widest following might "knock each other out" by dividing the vote, allowing little-known candidates to sneak through. The system puts voters in a difficult position of deciding whether they should vote for their favorite, who might be unlikely to win, or for one of the likely winners. The runoff system is also expensive; it requires the government to hold two full elections (ballots aren't free, after all).

It is almost too easy to point out the flaws in pure plurality rule. The most obvious weakness of the plurality rule is that the winner need not have widespread support. Suppose there are four candidates, $\{w, x, y, z\}$, running for mayor. The voters can be divided into groups according to their preferences. In Table 3.1, the preference orderings of the four groups are illustrated. Note that the largest group, group 4, has 30 members and their favorite is w. This example is designed to emphasize the fundamental problem that a minority, 30% of the population, is able to force its will

	Voters			
Groups	1	2	3	4
Number of members	20	24	26	30
Ranking				
First	z	y	x	w
Second	x	z	y	z
Third	y	x	z	x
Fourth	w	w	w	y

Table 3.1   Plurality election with four candidates

on the rest, because all the other voters believe that w is the *worst possible choice*! One of the most troublesome aspects of this example is that w would lose a head-to-head election against x (because x is preferred by groups 1, 2, and 3), y, or z.

Holding a runoff between the top two vote getters would help somewhat, because x would win in a head-to-head competition against w. However, the voters who support z would certainly protest. Note that groups 1, 2, and 4 perfer z to x. How in the world can such a universally unlikeable character as x win an election?

The plurality problem is a "real life" problem with serious consequences. When a candidate wins an election with less than 50% of the vote, there is always concern that the wrong candidate won. Consider, for example, the fact that, in 1998, voters in the state of Minnesota elected former professional wrestler Jesse "The Body" Ventura as their governor. Ventura won with 37% of the vote, outdistancing the Republican candidate Norm Coleman (34%) and the Democrat Hubert Humphrey (28%). There's no way to know for sure what would have happened if either Coleman or Humphrey had faced Ventura in a one-on-one contest, but there's considerable speculation that either would have defeated him.

Even if you don't care much about who is governor of Minnesota, you probably do care who wins the coveted Grammy Awards for the best songs and albums. The winners are selected by a plurality vote with five nominees on the ballot. There is often concern that the best performers do not win. Music reporter Robert Hilburn suggested they consider changing their voting procedure, noting "Looking over previous Grammy contests, it's easy to see where strong albums may have drawn enough votes from each other to let a compromise choice win. In 1985, two of the great albums of the decade—Bruce Springsteen's "Born in the U.S.A." and Prince's "Purple Rain"—went head to head in the best album category, allowing Lionel Richie's far less memorable "Can't Slow Down" to get more votes" (Hilburn, 2002). Clever voters might try to outsmart the system, voting for their second-ranked alternative to stop a weak candidate from winning. Dishonest voting is euphemistically called strategic voting in the literature.

The plurality problem has been known for centuries. It was a focus of concern in French academic circles in the late 1700s. Two extremely interesting characters are the philosopher/scientists Jean-Charles de Borda and the Marquis de Condorcet. Recall that it was a time of revolution, both in America and in France. The philosophy of democracy was becoming well accepted. Marie Jean Antoine Nicolas Caritat, Marquis de Condorcet (1743–1794), was an extremely influential philosopher of the Enlightenment, studying not only voting (de Condorcet, 1785), but also publishing on the rights of women, slavery, and free markets. He was highly placed in academic circles, an eager proponent of revolution against the King, and eventually elected to the legislature (and then later imprisoned by an opposing faction). He is credited with a comment that translates as "The apparent will of the plurality may in fact be the complete opposite of their true will" (cited in Mackenzie, 2000b). Condorcet proposed a system of voting in which alternatives were paired off for comparison. Traveling in the same circles as Condorcet was Borda (1733–1799), an explorer, soldier, and scholar whose study of physics and mathematics had wide-ranging impacts on science. He agreed with Condorcet on the shortcomings of the plurality system and also proposed a new election procedure based on rank order voting. His manuscript, which was written sometime between 1781 and 1784 (Saari, 1994), indicates that he had previously presented his results on June 16, 1770 (de Borda, 1781). It appears that Borda had the best of it in the eyes of the French, because after the French Revolution (and until Napoleon took over) the Borda procedure was used in French elections. The Borda count fell into disuse after that, only to be revived in the modern era for rankings in American sports. Variants are used in legislative elections in the Pacific Island countries of Kiribati and Nauru (Reilly, 2002).

The debate between Borda and Condorcet has framed research on elections for two centuries. In the next section we consider Borda's method.

EXERCISES 3.3

1. Is there any way to win a majority of the vote without also winning a plurality of the vote?

2. Suppose there are 50 shares of stock in a company and the stockholders are allowed to vote on company policy. Each stockholder is allowed to cast one vote per share owned in a private ballot box. No voter's ballot has his/her name on it. Does this system violate the criterion of neutrality or anonymity?

3.   This example has 50 voters and three alternatives.

	Voters		
Groups	1	2	3
Number of members	24	16	10
Ranking			
First	x	y	x
Second	y	z	z
Third	z	x	y

a)   Find the majority rule winner, if there is one.

b)   Calculate the plurality vote totals for the candidates, find the winner.

c)   Does the plurality winner have a majority of the votes?

d)   Confirm that a runoff election will choose the same winner.

e)   If the winner of the each type of election is compared one on one against each of the other candidates, is it a winner?

4.   This example has 50 voters and three alternatives.

	Voters		
Groups	1	2	3
Number of members	24	16	10
Ranking			
First	x	y	z
Second	y	z	y
Third	z	x	x

a)   Which candidate would win a plurality election?

b)   Does the plurality winner have a majority of the votes?

c)   If there is a runoff election between the top two vote getters, which candidate will be the winner?

d)   If the winner of the each type of election is compared one on one against each of the other candidates, is it a winner?

5.   This example has 100 voters and four alternatives.

	Voters			
Groups	1	2	3	4
Number of members	24	19	21	36
Ranking				
First	x	y	z	w
Second	w	z	x	x
Third	z	x	y	y
Fourth	y	w	w	z

a) Find the plurality winner.

b) Does the plurality winner have a majority of the votes?

c) If there is a runoff election between the top two vote getters, which candidate will win?

d) Suppose that after the plurality vote only one candidate is eliminated, and then another three-way plurality election is held. Will any of the candidates win a majority? If not, which candidate will win a runoff election between the top two vote getters?

6. In 2003, the voters of California were presented with an interesting choice. First, do you want to remove Governor Gray Davis? Second, if Davis is to be replaced, which of the following 150 candidates would you most prefer? California law states that if the number who answer yes to the first question is more than 50% of the votes cast, then the candidate who receives the most votes will become governor. About 54% of the voters wanted to remove Davis, and the plurality winner of the second vote was body builder and movie star Arnold Schwarzenegger (48.6%). Consider these questions about the 2003 California election.

a) Can we be sure that a majority (or even a plurality) prefer Schwarzenegger to Davis as governor?

b) The results indicated that 46% of the people voted to keep Davis, but they lost. What percentage of the vote must the replacement receive to take office? Does that seem sensible?

c) What incentives did voters have to vote strategically (untruthfully) in the first and second parts of the ballot?

7. In the American criminal justice system, the members of a jury must agree unanimously to reach a verdict. If the result is not unanimous when deliberations end, there is a "hung jury" and a mistrial is declared. Explain why this system does not satisfy the principle of strong monotonicity.

8. Kenneth May showed that a system based on majority rule (in which indifferent voters abstain) is equivalent to a system in which the voters are anonymous, the alternatives are undifferentiated, and the procedure is strongly and weakly monotonic. In May's treatment, the voters consider two alternatives, x and y, and then cast votes by selecting a value from the set $\{-1, 0, 1\}$. A vote of 0 indicates indifference, whereas 1 indicates a preference for x and -1 indicates a preference for y.

   a) Create a mathematical definition of majority rule, assuming that the indifferent voters are ignored.

   b) Use your definition to consider the following: the winner is x and then one voter who favored y decides she is indifferent. If you apply your procedure for majority rule to the new voter preference data, does the result change?

   c) Suppose there is a tie and one voter changes her mind to prefer x. Apply your procedure to find out who the winner will be.

   d) On the basis of your calculations in b and c of this question, do you believe your definition of majority rule is consistent with the Monotonicity Principle?

3.4   Rank-Order Voting: The Borda Count

If you are a fan of college football, you may be aware of the fact that (until 2006, at least) there has been no national tournament to answer the question "which team is the best." Instead, 100 (or so) college football teams are ranked at the end of the season, and 8 of the top teams are selected to play in four spotlighted games that are known as the Bowl Championship Series (BCS). The top two ranked teams play in the national championship game, and after that game, there is another series of votes intended to rank all the teams from top to bottom.

The formula used by the BCS to decide which teams should play in the national championship game has been a source of frustration and controversy. In 2001, the University of Nebraska was soundly defeated by the University of Colorado in the final regular season game. That loss prevented Nebraska from winning the Northern Division of the Big 12 conference, and it was not in the Conference championship game (which Colorado won). Nevertheless, the BCS scores had Nebraska ahead of Colorado and the Nebraska Cornhuskers were selected to face the University of Miami in the championship game. Nebraska was soundly defeated, giving grounds to the contention that the BCS had chosen the wrong team. At the end of the 2003 season, the University of Southern California was not selected for the championship game, even though the polls of writers and coaches placed USC in the top two. Decisions like that have led to a chorus of complaints over the years, summarized aptly by a student sports writer at the University of Kansas: division 1A college football uses "what may be the most complicated monstrosity on the planet" (Bauer, 2004) to rank the teams.

The people who run the BCS are not oblivious to the criticism of their methods of ranking teams, and every few years they make repairs in their voting system. They raise or lower the weight put on the votes cast by coaches, sports writers, and, strangely enough, *computers*! So far, at least, they have not strayed far from the same basic system. Individuals (coaches, writers, and computers) are asked to provide rankings of the top 25 teams. The many individual rankings are summed to build an overall ranking. This is a variant of the method that was proposed by Jean-Charles de Borda (de Borda, 1781).

3.4.1   The Borda Count

The Borda count, which we label *BC*, is a method of combining the rankings of many individual voters. Suppose there are k alternatives. The voters rank the alternatives and assign integer scores to them, beginning with 0 (for the worst) stepping up in integer units to $k - 1$ (for the best). The rankings assigned by the voters are summed, and the alternative with the largest Borda count is the winner. The Borda count generates a complete transitive social ordering of the alternatives.

In mathematical terms, each voter submits a vector of scores that are keyed to the list of candidates. If the candidates are $\{x, y, z\}$ and voter i submits the *Borda vector* $v_i = (0, 2, 1)$, then the voter believes y is the best, z is second, and x is the worst. The Borda count earned by a candidate j is the sum of rankings assigned by the voters, $\sum_{i=1}^{n} v_{ij}$. The notation $x \succ_{BC} y$ indicates that x defeats y in the Borda count. Although we do not concern ourselves with ties in this chapter, we should mention that voters are allowed to register indifference by assigning two alternatives the same value. For example, $(0.5, 0.5, 2)$ indicates that x and y are equally desirable (one vote is split between them), and both are less appealing than z.

The Borda count works sensibly in many situations. For example, if all the voters agree about the ranking for all the alternatives, it is easy to see that the Borda count gives a meaningful ranking. The rankings assigned by the three voters are presented in Table 3.2A. The *Borda table*, shown in Table 3.2B, illustrates the numerical vectors that represent the Borda votes. This demonstrates that the "sum of ranks" procedure used in the BCS and other poll-style ranking systems is the same as the Borda count. It is, of course, painfully obvious that x should win. It is the most-preferred alternative for all the voters and it has the highest Borda count.

A system of this sort has been the basis of many ranking systems used in sports and entertainment. Although there are bitter squabbles about the outcomes of these ranking procedures, the commentary in the public media seldom considers the possibility that there is a fundamental problem in the basic idea of the Borda count itself. Very peculiar—even alarming—things can happen with the Borda count.

A. Voter preference profile

	Voter rankings		
Voter	1	2	3
Rankings			
First	x	x	x
Second	y	y	y
Third	z	z	z

B. Borda count table

	Borda votes			Borda count
Voter	1	2	3	
Alternatives				
x	2	2	2	6
y	1	1	1	3
z	0	0	0	0

Table 3.2 Borda count with homogeneous voters

3.4.2 A Paradox in the Borda Procedure

The problem that we now introduce has sometimes been called the "inverted-order paradox" or the "winner turns loser paradox." It is a famous example that has been widely investigated (Fishburn, 1974; Riker, 1982, p. 82). There are seven voters, {1, 2, 3, 4, 5, 6, 7}, and four alternatives, {w, x, y, z}. In Table 3.3, the columns represent the Borda voting vectors assigned by the voters. The winner, the candidate with the largest Borda count, is y. The overall ranking, from top to bottom, is $y \succ_{BC} x \succ_{BC} w \succ_{BC} z$. This is a strength of the Borda count. It generates a transitive ordering. Nevertheless, in the eyes of its critics, the ordering is complete nonsense.

To bring some life into this discussion, suppose that the alternatives w, x, y, and z are bands competing the prestigious Grammy award. After the votes have been collected, an investigative reporter exposes the fact that group z is composed of frauds who lip-sync their performances.[1] After z has been exposed, the Grammy managers declare z ineligible. It would seem logical to expect that if the ordering was $y \succ_{BC} x \succ$

---

[1] A group later shown to have been lip-syncing, *Milli Vanilli*, actually did win a Grammy award for Best New Artist after their success in 1989. That award was revoked, and Milli Vanilli was removed from the official list of award-winning performers.

	Borda votes							Borda count
Voter	1	2	3	4	5	6	7	
Alternatives								
w	3	3	0	0	1	1	3	11
x	2	2	3	3	0	0	2	12
y	1	1	2	2	3	3	1	13
z	0	0	1	1	2	2	0	6

Table 3.3 Borda count with four alternatives

$_{\text{BC}}$ $w \succ_{\text{BC}} z$ and we rule the last-place candidate ineligible, then the new ordering among the remaining alternatives should simply be $y \succ_{\text{BC}} x \succ_{\text{BC}} w$. Band z was a loser. The winner should still be y, shouldn't it?

Sadly enough for the proponents of the Borda count, the answer is no. Assuming that y would still win was a terrible mistake caused by a fundamental flaw in the Borda count. We can recalculate the votes, supposing either that z is completely eliminated from consideration or that it is forced into last place on all the ballots (the result is the same, either way). Let's begin by simply moving z to the bottom of each voter's ranking. This will ensure that z cannot be the winner because it will finish in last place. The alternatives that were ranked below z are shifted up one notch, as indicated in Table 3.4. After moving z to the bottom of each ballot, the Borda count indicates that w should be the winner. The ranking from top to bottom is $w \succ_{\text{BC}} x \succ_{\text{BC}} y \succ_{\text{BC}} z$. Compare that with the previous result and a truly astonishing

	Borda votes							Borda count
Voter	1	2	3	4	5	6	7	
Alternatives								
w	3	3	1	1	2	2	3	15
x	2	2	3	3	1	1	2	14
y	1	1	2	2	3	3	1	13
z	0	0	0	0	0	0	0	0

Table 3.4 Borda count after making z ineligible

outcome is revealed: *The ordering of the top three alternatives is exactly reversed.* Instead of winning, y now finishes in third place.

The befuddling aspect of this result is that the Borda ranking of w and x is influenced by the placement of the alternative z. If we are completely focused on the comparison of w and x, then z is an *irrelevant alternative*. When the irrelevant alternative affects the outcome, then the decision-making procedure is said to violate the principle of *Independence From Irrelevant Alternatives*, which holds that the social ranking of two alternatives should not be influenced by the placement of other alternatives in the ballots of the voters.

A story might persuade readers that a voting procedure ought to respect the principle of Independence From Irrelevant Alternatives. Suppose your friend Daffy and his family have received a gift certificate for a new car, either a Toyota or a Honda. When the Daffy family is focused on those two alternatives, the desirability of all other makes should be irrelevant. Suppose you are discussing this in the car dealer's parking lot and Daffy says out loud, "We prefer the Toyota to Honda. We will go in to pick that one." You figure all the work is done, so you go home. Later, Daffy drives up to your house in a shiny new Honda. You say, "What happened? You preferred Toyota!" Daffy says, "While we were in line, my son heard that Cadillacs have great durability. So we changed from Toyota to Honda!" Could the outcome be more ridiculous? It cannot possibly make sense to have the choice between a Honda and a Toyota depend on the road performance of a Cadillac. And yet, that can happen with the Borda count.

In the example illustrated in Tables 3.3 and 3.4, the terminology says z is an irrelevant alternative, but it is not, in fact, irrelevant when the Borda count is used. That is the sense in which the Borda count violates the principle of Independence From Irrelevant Alternatives. Although commentators on amateur figure skating competitions do not know the terminology, they have certainly puzzled over it many times because skating uses a Borda-style ranking system. In the 1995 World Figure Skating Championships, Michelle Kwan's surprisingly good performance put her in fourth place but caused a reversal in the ranking of the second and third ranked skaters who had already finished before Kwan took the ice (Mackenzie, 2000a).

Another variant of this story begins with three alternatives. In Table 3.5, the three alternatives are w, x and y. The Borda winner is w. Now, suppose that someone who is in charge of the election desperately wants y to win. That someone is so desperate, in fact, that he is willing to sponsor the entry of a new candidate into the race. The new candidate, z, is chosen very carefully. When z is in the race, the preferences of the voters are the same ones we considered in Table 3.3. After adding in the losing candidate z, the Borda count selects y, just as the election manager wants. This displays the truly insidious nature of the violation of the principle of irrelevant alternatives. Not only is the decision illogical, but now it is prone to electoral manipulation. Without having precise information about that tastes of the voters, it might be difficult to slide in just the right candidate to bring about the desired outcome. One might mistakenly make the outcome even less desirable.

Voter	Borda votes							Borda count
Alternatives	1	2	3	4	5	6	7	
w	2	2	0	0	1	1	2	8
x	1	1	2	2	0	0	1	7
y	0	0	1	1	2	2	0	4

Table 3.5   Borda count with three alternatives

3.4.3   Another Paradox: The Borda Winner Is a Loser?

There is another closely related (equally serious) problem with the Borda count. The Borda count can produce a winner that would lose in a head-to-head election against one of the other alternatives. In Table 3.3, the Borda winner is y, but in a head-to-head contest between y and x, x would win because it is preferred to y by voters {1, 2, 3, 4, 7}. This result is not necessarily a sign of trouble in the eyes of the advocates of the Borda count. The Borda method, they argue, averages out the extreme opinions, so that the winner is somehow "more representative" of the overall opinions of the voters.

The problem is actually more serious. When he proposed his voting system, Jean-Charles de Borda drew the attention of a fellow French academic, the Marquis de Condorcet. Condorcet pointed out what he thought was a crippling weakness in Borda's method. Consider Table 3.6, an example in which there are five voters and four alternatives. The Borda winner is x, but x is defeated in a head-to-head contest with w. In fact, w would defeat each of the other alternatives in a head-to-head contest. Alternative w is preferred to x by voters {1, 2, 3}, and to y by voters {1, 2, 4} and to z by voters {1, 2, 5}. In Condorcet's honor, we call an alternative that can win a pairwise (head-to-head) contest against each of the other alternatives the *Condorcet*

Voter	Borda votes					Borda count
Alternatives	1	2	3	4	5	
w	3	3	1	1	1	9
x	2	2	0	3	3	10
y	1	0	2	0	2	5
z	0	1	3	2	0	6

Table 3.6   Condorcet winner (w) rejected by Borda count

winner. A voting procedure that does not select the Condorcet winner is said to violate the *Condorcet criterion.*

3.4.4   Inside the Guts of the Borda Count

The ranking created by the Borda count is vulnerable to manipulation by the addition or subtraction of candidates. By working on a few examples, one can gain some good working knowledge of what kinds of changes make a difference. One of the clearest examples is found by comparing an election with two candidates against an election with three candidates. Suppose there are 100 voters and they are divided into two groups. There are 60 voters in group 1 and they prefer x to y. Formally speaking,

$$\text{Group 1: } x \succ_i y \text{ for } i \in \{1, 2,..., 60\}$$

There are 40 voters on the other side of the issue:

$$\text{Group 2: } y \succ_i x \text{ for } i \in \{61, 62,... 100\}.$$

The Borda count is summarized in Table 3.7. Because group 1 has the most voters, its favorite is the winner.

Next, suppose that a third alternative is added, and it is very similar to the loser, y. We want to insert this new alternative into the preferences of the voters so that z is always right next to y. Because z is always grouped together with y, it is sometimes called a *clone.* (One standard for voting procedures is that their outcomes should not be changed by the addition of redundant alternatives (clones). This example is intended to show that the Borda count is not "cloneproof.") There are four possible orderings. The first two new orderings are found by placing z about y in the tastes of group 1:

$$\text{Group 1a: } x \succ_i y \succ_i z$$
$$\text{Group 1b: } x \succ_i z \succ_i y$$

	Borda votes		Borda count
	1	2	
Groups	1	2	
Number of members	60	40	
Alternatives			
x	1	0	$60 \cdot 1 = 60$
y	0	1	$40 \cdot 1 = 40$

Table 3.7   Borda count with two alternatives

The last two are obtained by inserting z into the preferences of group 2:

$$\text{Group 2a: } z \succ_i y \succ_i x$$
$$\text{Group 2b: } y \succ_i z \succ_i x$$

By inserting z, which is very similar to y in the eyes of the voters, we thus obtain preferences for four groups. The relative sizes of these groups are referred to as n_{1a}, n_{1b}, n_{2a}, and n_{2b} in Table 3.8.

Suppose that the original groups are split exactly in half, so $n_{1a} = 30$, $n_{1b} = 30$, $n_{2a} = 20$, and $n_{2b} = 20$. With that even division within the two groups, the Borda count is 120, 90, 90 for x, y, and z, respectively. Because y is similar to z in the eyes of the voters, the result seems plausible because y and z are tied in the Borda count. The original winner, the most favored alternative of the majority group, still wins.

We can make some magic by fiddling with the sizes of the subgroups. Keep in mind that there are always 60 voters for whom x is the most attractive alternative and there are always 40 for whom x is the least attractive. The only manipulation that we consider is the subdivision of the original two groups. (If you use a computer spreadsheet, it is pretty easy to consider lots of conjectures.) For most of the examples that you try, the winner is x, but there are some exceptions. The exceptions are found when there are many voters who have y preferred to z. That is, if you increase the values of n_{1a} and n_{2b} enough, then the Borda winner changes from x to y. If the groups are divided $n_{1a} = 50$, $n_{1b} = 10$, $n_{2a} = 5$, and $n_{2b} = 35$, then the Borda counts for x, y, and z are 120, 125, and 105. Even though x is still the first-ranked alternative for 60 of 100 voters, the alternative y is the Borda winner.

The algebra of the situation is enlightening. Alternative y will defeat x (i.e., have a higher Borda count) if

$$1 \cdot n_{1a} + 1 \cdot n_{2a} + 2 \cdot n_{2b} > 2 \cdot n_{1a} + 2 \cdot n_{1b}$$

which is easily simplified:

$$n_{1a} + 2 \cdot n_{1b} - n_{2a} - 2 \cdot n_{2b} < 0$$

Groups	Borda votes				Borda count
	1	2	3	4	Borda count
Number of members	n_{1a}	n_{1b}	n_{2a}	n_{2b}	
Alternatives					
x	2	2	0	0	$2 \cdot n_{1a} + 2 \cdot n_{1b}$
y	1	0	1	2	$1 \cdot n_{1a} + 1 \cdot n_{2a} + 2 \cdot n_{2b}$
z	0	1	2	1	$1 \cdot n_{1b} + 2 \cdot n_{2a} + 1 \cdot n_{2b}$

Table 3.8   Add one alternative to Table 3.7

Keeping in mind that $0 \leq n_{1b} \leq 60$ and $n_{1b} = 60 - n_{1a}$ as well as $0 \leq n_{2a} \leq 40$ and $n_{2a} = 40 - n_{2b}$, the Borda count for y will be superior if

$$n_{1a} + n_{2b} \geq 80$$

In the above example, where we set $n_{1a} = 50$ and $n_{2b} = 35$, this inequality was satisfied and so y was the winner.

Recall that the premise of this example is that the new alternative, z, is similar to (a clone of) y, and z is inserted into the preferences in the spot immediately before or after y. For the introduction of z to cause a change in the Borda count that benefits y, it is vital that y be more desirable than z in the eyes of many (at least 80) of the voters. One might criticize this example by pointing out that y and z are not truly similar because a vast majority of the voters see y as more desirable than z, and if they were really similar, the split would be more even. This argument, although interesting, does not rehabilitate the Borda count. Rather, it simply refocuses our attention on the problem of irrelevant alternatives. If the overall ranking of x against y depends on the question of whether y is more desirable than z, then one ought to be cautious about the result from a Borda count. If x loses, its supporters must surely feel abused because the position of their favorite against y is not changed in the transition from Table 3.7 to Table 3.8. On the other hand, if y is the loser and x is the winner, then the supporters of y are unhappy and feel unjustly treated because their loss was caused by the high placement of z in the rankings of some voters. It just doesn't seem right. It seems this kind of illogic is fundamental in procedures based on the Borda method of voting. And, in case you were wondering, that means the BCS football ranking system will likely produce exasperating results.

3.4.5   Digression on the Use of Cardinal Preferences

The element of the Borda procedure that is most often thought to be the culprit is the simple method in which preferences are registered. The preferences of the voters are registered as integer-valued rankings. If a voter believes that x is much better (or just a little better) than y and y is much (or just a little bit better) than z, the ranking will be submitted as 1-2-3 and the magnitude of preference is ignored. A proponent of cardinal preferences would rather have us figure out a way to precisely measure these differences so that the voter could submit a vote like 1-1.8-2.0 to signify the fact that the first one is the best and the other two are far worse. In the advanced literature on social choice theory, there are in fact some highly prestigious authors, such as Nobel Prize Winners John Nash (1950) and John Harsanyi (1955), who advocated decision making based on these fine-grained evaluations. John Nash, the game theorist whose life story was the basis of the movie *A Beautiful Mind*, was probably the first to contend that these cardinal scores should be collected and *multiplied* together to create a social ranking. In his honor, Riker calls this the Nash method (Riker, 1982).

The use of cardinal scores has intuitive appeal. Harsanyi offers a rigorous mathematical argument in favor of this system, contending that it optimizes the welfare of the community in a utilitarian sense. That may be, but, as Riker demonstrates (1982, pp. 110–111), the problem of irrelevant alternatives remains. In fact, the problem may be more severe in the Nash method than in the Borda method. In the Borda method, the fact that voters are restricted to casting votes in integer values limits the extent to which the placement of z might influence the comparison of x and y. In a cardinal election, the votes can range across a continuum, seemingly opening up a much larger range of outcomes that are influenced by irrelevant alternatives.

EXERCISES 3.4

1. Consider a two-candidate race for the presidency of a social club. There are 135 members of the club and 81 prefer *Fred* to *Barney*, whereas the remaining members are of the opposite opinion. Write down the Borda vote vectors that would be cast and find the winner of the Borda count. Is the same candidate going to win a majority rule election?

2. This example has 65 voters and three alternatives.

	Voters		
Groups	1	2	3
Number of members	22	12	31
Ranking			
First	x	y	z
Second	y	x	y
Third	z	z	x

a) Which candidate would win a plurality election?

b) What would be the results of a majority/runoff election?

c) What would be the Borda vector for each group of voters?

d) Create a Borda count table. What would be the Borda count for each candidate?

e) Is there a Condorcet winner in this example?

3. This is a set of preferences for a society of 1000 voters with four alternatives

Groups						
	1	2	3	4	5	6
Number of members	200	150	150	300	100	100
Ranking						
First	z	y	x	w	y	x
Second	x	z	y	z	x	y
Third	y	x	z	x	w	z
Fourth	w	w	w	y	z	w

a) Is there a Condorcet winner? If so, which candidate?

b) Create a Borda count table. Which candidate would win according to the Borda count?

c) Delete candidate w and recalculate the Borda count.

d) Does the Condorcet winner change after deleting w?

4. Calculate the Borda count for the following table. Then add an alternative that finishes in last place in the Borda count but also causes the ranking of the top-placed alternatives to change.

Voter			Borda votes					Borda count
	1	2	3	4	5	6	7	
Alternatives								
w	2	2	0	0	1	1	2	?
x	1	1	2	2	0	0	1	?
y	0	0	1	1	2	2	0	?

5. Condorcet's original example that was used to criticize the Borda count involved an election with candidates named *Jacques*, *Pierre*, and *Paul*.

Groups			Borda votes			
	1	2	3	4	5	6
Number of voters	30	1	10	29	10	1
Alternative						
Pierre	2	2	1	1	0	0
Paul	1	0	0	2	2	1
Jacques	0	1	2	0	1	2

a) Find the Condorcet winner.

b) Find the Borda winner.

c) How many ballots would you have to change to make the outcome of the two procedures match up?

3.5   Sequential Pairwise Comparisons

The critics of the Bowl Championship Series often contend that college football should adopt a tournament format to select a national champion. A tournament is used in many other college sports as well as professional baseball, basketball, and football. Would a tournament solve the controversy over "who's number one?" There are good reasons to be skeptical. A tournament might increase advertising revenue, but we doubt it would put an end to questions about whether the best team was actually ranked number one at the end.

3.5.1   A Single Elimination Tournament

The teams are eliminated in a series of games, the final of which would be the national championship game. A hypothetical bracket for such a tournament is presented in Figure 3.1. The teams that are defeated at each stage are eliminated, and the winners advance to play each other.

The people who want a Division IA college football tournament usually claim that this is the best way to select the "true champion." In the hypothetical example in Figure 3.1, the University of Sasnak wins. Whether or not Sasnak is greeted as the

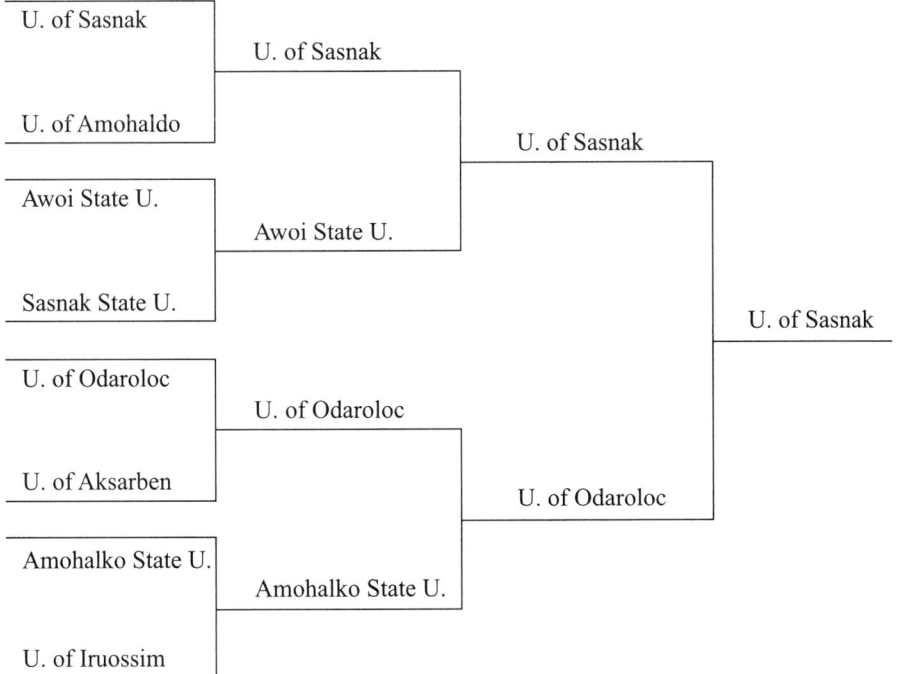

Figure 3.1   *Hypothetical football tournament.*

"true champion" may be open to question, however. Suppose, for example, that Sasnak could be defeated by Sasnak State, Aksarben, Amohalko State, or Iruossim. Those other teams are not given their "fair chance" to prove their superiority because the tournament structure governs their opportunities. If even one team can defeat the eventual winner of the tournament, it seems we are going to be forever debating what would have happened. The tournament does not so much solve the problem of deciding "which team is best" as it does create a new set of arguments for people in bar rooms and sports talk shows.

The tournament concept, it turns out, is widely used, not just for sporting events but for political decision making as well. For example, in American elections, the candidates of the top two political parties face each other in the general election (the "championship game"). Those candidates are the winners of earlier contests. Like the winners of sporting tournaments, these winners are no less open to after-the-fact challenges from contenders who did not get their fair chance.

There is one case in which a tournament structure produces an unequivocal certain winner. If there is a Condorcet winner, one candidate (or team) that defeats each of the others in a one-on-one competition, then that winner will be the only survivor of the tournament.

3.5.2   Dominated Winner Paradox

Like the Borda count, the majority rule tournament format can produce some truly bizarre outcomes. The tournament never rejects a Condorcet winner, if there is one, but interesting things can still happen. Perhaps the most well known is the "dominated winner paradox." It is possible that the winner of a tournament can be *unanimously defeated* by another alternative. We mean to say not just that the tournament picks the second-best or third-best, but rather that the tournament winner is unanimously considered inferior. The preferences in Table 3.9 are used to illustrate this paradox. There are three voters and four alternatives, w, x, y, and z. In the tourna-

	Voter preferences		
Voters	1	2	3
Rankings			
First	w	y	x
Second	x	w	z
Third	z	x	y
Fourth	y	z	w
Sequence of votes	$w \succ_M x$	$y \succ_M w$	$z \succ_M y$

Table 3.9   Dominated winner paradox

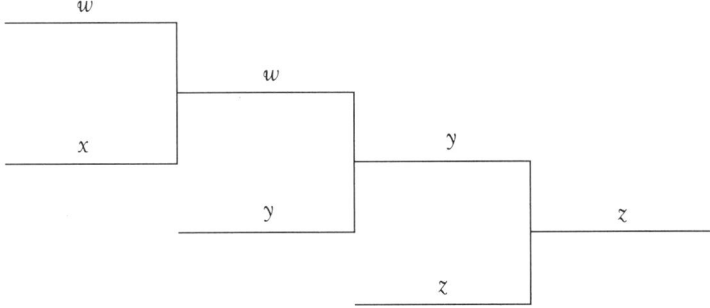

Figure 3.2   *Tournament structure for the dominated winner paradox.*

ment structure that is drawn in Figure 3.2, there is a sequence of three votes. First, alternative x is defeated by w. Then, in the second stage, the winner w is defeated by y, and then the final vote pits y against z and z wins. Note that the first outcome to be eliminated, x, is preferred to z by every single voter.[2]

3.5.3   The Intransitivity of Majority Rule

In the first section of this chapter, we drew the reader's attention to the concept of transitivity. Now that concept comes to the forefront again. There are majority rule examples in which $x \succ_M y$ and $y \succ_M z$ but (and here's nontransitive part) $z \succ_M x$. Social comparisons sometimes lack the fundamental properties of reason and logic summarized by the principle of transitivity that is present in the tastes of voters.

Consider an example with three candidates (two Democrats, one Republican) and three groups of voters. The Democrats, named *Thing 1* and *Thing 2*, are pitted in a tightly contested electoral campaign, and the winner will face the Republican candidate, named *Cat in the Hat*. Let's suppose there are three equally sized groups of voters. The transitive orderings of the voters are presented in Table 3.10. Readers will note that this kind of table is slightly different from the one used for the Borda count, but the information that it contains is the same.

It is especially important to note that the preferences of the voters are *transitive*. Consider the preferences in Table 3.10. For group 1 we are told that *Thing 1* is preferred to *Thing 2* and *Thing 2* is preferred to *Cat in the Hat*. Then transitivity of preference implies we are safe in concluding that *Thing 1* is preferred to the *Cat in the Hat* by members of group 1.

The tournament begins with *Thing 1* and *Thing 2* facing each other and all three groups of voters are allowed to have their say. Because *Thing 1* is preferred to *Thing 2*

---

[2]This is called the "dominated winner paradox" because, in the field of cooperative game theory, one alternative is said to dominate another if every participant prefers it to the other.

	Voter groups		
	1	2	3
Rankings			
First	*Thing 1*	*Cat in the Hat*	*Thing 2*
Second	*Thing 2*	*Thing 1*	*Cat in the Hat*
Third	Cat in the Hat	*Thing 2*	*Thing 1*

Table 3.10   Preferences for three candidates

by groups 1 and 2, *Thing 1* is the winner of that stage. The final stage pits *Thing 1* against *Cat in the Hat*, and with the backing of groups 2 and 3, *Cat in the Hat* wins.

If majority rule were transitive, then we could conclude that *Cat in the Hat* is socially preferred to *Thing 2*. It is a startling and truly paradoxical fact that majority rule does not support that prediction. Majority rule does not obey the property of transitivity.

Imagine what would happen if, in the time-honored tradition of politicians, *Thing 2* hires a large team of lawyers who appeal the case to the highest court. The court might order another election pitting *Thing 2* against *Cat in the Hat*. The *Cat in the Hat* should do everything he can to avoid another vote. Note that groups 1 and 3 prefer *Thing 2* to *Cat in the Hat*, so *Thing 2* would win that final electoral competition.

Even though the individual voter preferences are transitive, the majority rule ordering is not! This peculiar finding, sometimes called the *Voter's Paradox*, means that where we expect to find logic and reason—$x \succ_M y$, $y \succ_M z$, and $x \succ_M z$—instead we find nonsense and irrationality—$x \succ_M y$, $y \succ_M z$, and $z \succ_M x$. We could keep voting forever, with each winner being defeated in turn. This is called a *voting cycle*. A person who is intransitive lacks the fundamental properties of reason and logic and would be incapable of making decisions and managing personal affairs. Should a society that is intransitive be viewed in the same harsh light? For centuries, political scientists have been locked in debate over the issue.

Despite these problems, majority rule is very widely used in legislatures. There are some arguments that can be advanced in favor of the majority rule method. By far the most important justification for the use of majority rule is offered by May's theorem, which we have already discussed.

The second reason why majority rule is still widely used, even though cycles are possible, is that cycles are thought to be rare. If there are only three voters with three alternatives and we ignore the possibility that voters may be indifferent, then we can make a list of all possible "societies." There are 216 possible three-voter societies, and a cycle can arise in 12, or approximately 5.6% (Johnson, 1998, p. 23). Many computer simulation studies have been done to try to find out how likely a cycle is to arise when the numbers of voters and alternatives are increased. A review of several studies led Riker (1982) to contend that the probability of cycles rises dramatically (near 1.0) as the numbers of alternatives and voters increase, whereas Jones et al. (1995) contended that if voter indifference is taken into account, then the probability of cycles is not quite so high.

If cycles are possible in the real world, why don't we observe them more often? Quite simply, we do not use voting procedures that are designed to discover them. Election procedures are generally written so that candidates are eliminated and are not given a second chance. Voting in legislatures is tightly controlled by party leaders who refuse to let the members propose frivolous alternatives in a "fishing expedition." In his discussion of the American Constitutional Convention of 1787, Riker tells the story of a cycle that was revealed when the delegates were attempting to formalize the system for presidential elections (1986, pp. 46–47). The cycle caused confusion and uncertainty, and eventually the matter was delegated to a subcommittee that was trusted to write up something good. That's how the Electoral College was created.

EXERCISES 3.5

1. The candidates in an electoral tournament are *Hamilton, Joe, Frank,* and *Reynolds.* In the first stage, *Reynolds* is paired off against *Frank,* and *Joe* faces off against *Hamilton.* The winners of those two contests compete to decide the overall winner. Draw a figure that represents this tournament.

2. Here is a preference profile for the voters who are polled in the tournament described in the previous question.

	Voters		
Groups	1	2	3
Number of members	22	12	31
Ranking			
First	Hamilton	Joe	Frank
Second	Frank	Hamilton	Reynolds
Third	Joe	Reynolds	Joe
Fourth	Reynolds	Frank	Hamilton

a) Calculate the winning candidate at each stage.

b) Check to find out if there is a Condorcet winner.

c) Is it possible for one of the losers in the first stage to defeat the eventual winner of this tournament?

d) Figure out if the winner of your tournament is the same as the result of the plurality, majority/runoff, and Borda count procedures.

3. Here is a preference profile:

	Voters		
Groups	1	2	3
Number of members	33	33	33
Ranking			
First	x	y	z
Second	y	x	x
Third	z	z	y

a) Confirm that there is a Condorcet winner.

b) Suppose there is a tournament in which x and y face off against each other, and then z challenges the winner. Does the Condorcet winner also win the tournament?

c) Can you think of any way in which to design the tournament so that the Condorcet winner does not win the final contest of the tournament?

4. Consider the following preferences.

	Voters		
Groups	1	2	3
Number of members	18	16	21
Ranking			
First	x	y	z
Second	y	x	y
Third	z	z	x

a) Design a single elimination tournament and find the winner.

b) Is there any way to redesign your tournament so that a different alternative will win?

5. In the following table, we provide preferences for 7 of 11 voters.

	Voters					
Groups	1	2	3	4	5	6
Number of members	3	3	2	1	1	1
Ranking						
First	z	y	x			
Second	x	z	w			
Third	y	w	z			
Fourth	w	x	y			

a) Fill in the preferences for the last three columns in such a way that there is a pairwise voting cycle in which $x \succ_P y$, $y \succ_P z$, and $z \succ_P x$. You can insert any transitive rankings for groups 4, 5, and 6.

b) Choose your favorite letter from the set $\{x, y, z\}$. Design a single-elimination tournament in which your selected alternative wins.

3.6   Condorcet Methods: The Round Robin Tournament

On the basis of the preceding analysis, the reader should believe that the following claims are correct.

1. If there is a Condorcet winner, then a single-elimination tournament format selects that alternative.

2. If there is no Condorcet winner, the tournament winner is determined by the pairings of the alternatives.

The presiding officer of a town council might exercise the power to set the agenda to the advantage of some community groups over others. There's a charming essay about an economist and a lawyer who studied social choice theory and then used it to hornswaggle the voters in a club that purchased airplanes for recreational use (Riker, 1986).

To address this problem, Condorcet's approach was to search for a method of voting that would give a meaningful result when all the pairwise comparisons are considered. Condorcet suggested that we collect enough information from voters so that we can hold (what is now called) a "round robin" tournament. In a round robin, each alternative faces each of the others in a head-to-head competition.

The *Condorcet criterion* states that if there is a Condorcet winner—one alternative can defeat each of the others head-on—then it should win. If there is no Condorcet winner, then the problem is to find a way to summarize the pairwise information and to choose or shape the results into a ranking. Duncan Black, a pioneer of modern social choice research, suggested the following: "The reasons may not seem so overwhelmingly convincing, but we are moving in a region where all considerations are tenuous and fine-spun; and the claims of the Condorcet criterion to rightness seem to us much stronger than those of any other" (Black, 1958). In the time since Condorcet, many different schemes have been proposed to summarize the outcomes of pairwise comparisons. Black suggested the use of a Borda count. We consider just a few of the many interesting proposals.

3.6.1   Searching for an Unbeatable Set of Alternatives

Here's an obvious starting point: eliminate undesirable alternatives from consideration. Suppose the alternatives under consideration are $\{u, v, w, x, y, z\}$. In Figure 3.3, the relationships among the alternatives are represented by arrows. In this type of graph, an arrow from x to y means that x defeats y is a pairwise competition. In Figure 3.3a, note that u, y, and x, the ones that are inside the dotted line, have a special quality. Each of them can defeat each of the others that are outside the dotted line in a head-to-head race. It seems clear that the winner must not be drawn from $\{v, w, z\}$. Each of these can be defeated by each of the proposals in $\{u, x, y\}$. The *Smith criterion* is based on the idea that if we can spot some "rejects" in this way, then we ought to make sure they don't win.

Formally, the *Smith set* is defined as the smaller of two sets:

1.   The set of all alternatives, X.

2.   A subset $A \subset X$ such that each member of A can defeat every member of X that is not in A, which we call $B = X - A$. Formally, the Smith criterion states if $x \in A$ and $y \in B = X - A$, $x \succ_P y$.

Note we are using the plurality rule notation here, $x \succ_P y$ [see (3.6)]. Indifferent voters are ignored. As we have seen in Figure 3.3a, the Smith set may include a voting cycle, and that's why some authors refer to it as the "top cycle set." The top cycle includes only the "best of the best." The rest of the alternatives should be rejected.

One major shortcoming of the Smith criterion is illustrated in Figure 3.3b. We have turned around a few arrows, so there is no subset of alternatives in which each can defeat the rest. In such a case, the Smith set includes all alternatives. As a result, the Smith criterion sheds no light at all on the problem. The Smith criterion can be useful in ruling out alternatives in some examples, but not in too many.

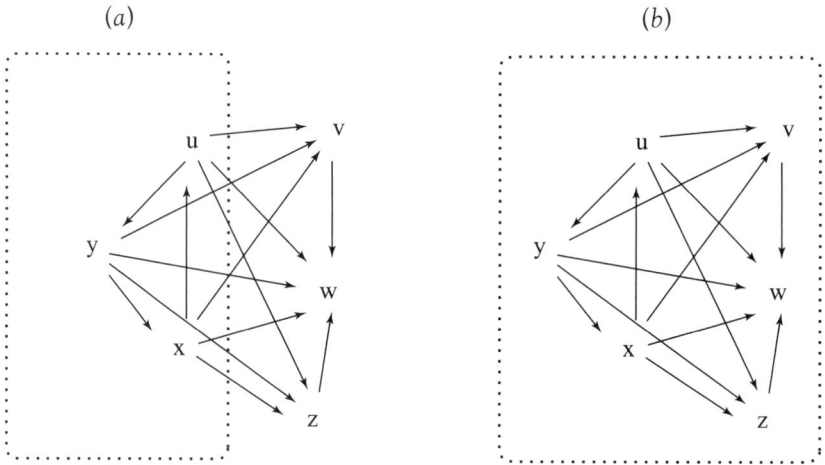

Figure 3.3   *The Smith set. (a) The Smith set is a "top cycle." (b) The Smith set includes all alternatives.*

3.6.2 The Win–Loss Record

A second simple approach is to choose a winner on the basis of the win–loss records. This is often called the Copeland rule. We simply subtract the number of losses from the number of wins for each alternative and then rank the alternatives. If one reads the sports page during professional baseball or football seasons, the win–loss record is a very familiar concept. Although simple in principle, this procedure has many shortcomings. Most importantly, if there is a voting cycle, then there are likely to be many ties in the win–loss column. Each alternative has one win and one loss in the pairwise comparisons, so none can be distinguished from the others. Another problem with this approach is its sensitivity to the introduction of clones. One could pad the number of wins for an alternative by making copies of the alternatives that it can defeat. (For Three Stooges fans, if Curly is a loser, then nominate Moe and Larry as well so that the alternatives that defeat Curly also defeat Moe and Larry.)

3.6.3 Aggregated Pairwise Voting: The Borda Count Strikes Back!

The aggregated pairwise vote is a method of evaluating a tournament. The alternatives accumulate votes in their head-to-head contests and the one with the most votes wins. It creates a transitive overall ranking, even if there is a pairwise majority rule cycle.

Suppose the voters are $N = \{1, 2, ..., n\}$ and that none of them is indifferent between any of the alternatives. Recall that the number of voters who prefer x to y can be represented as $|\{i \in N : x \succ_i y\}|$. The aggregated vote for alternative x is the sum of the support it receives against y and z in pairwise contests:

$$|\{i \in N : x \succ_i y\}| + |\{i \in N : x \succ_i z\}|$$

The aggregated pairwise vote has a number of interesting properties. Readers should easily convince themselves that the following claims are true:

1. A Condorcet winner (one who is undefeated by each of the others) is never ranked last by the aggregated pairwise vote.

2. A Condorcet loser (one that is unable to defeat any of the others) is never ranked first by the aggregated pairwise vote.

The aggregated pairwise vote is never afflicted by the "dominated winner paradox" that was discussed above. Although the Condorcet winner does not always come out on top in an aggregated pairwise vote, that candidate does not suffer the indignity of a last place finish either.

It turns out that these two properties of the aggregated pairwise vote have a powerful implication: Only a restricted number of paradoxical voting outcomes is possible. With two alternatives, the aggregated pairwise system is really just plurality rule, so we might write either $x \succ_{AP} y$ or $x \succ_P y$. If the pairwise result is $x \succ_{AP} y$, then a supporter of Condorcet's view of the problem would expect that the social ranking should be $x \succ_{AP} y \succ_{AP} z$ or $z \succ_{AP} x \succ_{AP} y$. As long as the procedure applied to all three alternatives has x preferred to y, then everything makes sense. On the other hand, there is trouble if we observe social rankings like $y \succ_{AP} x \succ_{AP} z$ or $z \succ_{AP} y \succ_{AP} x$.

Mathematician Donald Saari offers a way to catalogue these so-called anomalies. For a given set of voters (a preference profile), any voting procedure can be applied to rank the pairs, {x, y}, {y, z}, {x, z}, and also give an ordering of all three alternatives, {x, y, z}. Because the pairwise comparison is really just a plurality vote, let's use P as the subscript for pairwise comparisons and AP for the comparisons that take into account three or more alternatives. Saari uses the term *word* to refer to a combined set of pairwise and aggregated pair-wise rankings, such as

$$\text{word 1.} x \succ_P y, \quad y \succ_P z, \quad x \succ_P z, \quad x \succ_{AP} y \succ_{AP} z$$

or

$$\text{word 2.} x \succ_P y, \quad y \succeq_P z, \quad x \succ_P z, \quad z \succ_{AP} y \succ_{AP} x$$

There are 351 words possible with three alternatives (hint: $3 \cdot 3 \cdot 3 \cdot 13 = 351$). In word 1, there is no anomaly because each of the pairwise plurality decisions is the same as the ordering implied by the aggregated pairwise votes. On the other hand, word 2 is very anomalous. The pairwise comparisons are at odds with the overall ranking. In fact, recalling the first claim above, readers should notice that word 2 is impossible. The aggregated pairwise vote could not assign last place to the Condorcet winner, x.

Saari observes that only 135 of the 351 words are actually possible with the aggregated pairwise vote applied to three alternatives (1994, p. 186). The driving force behind that result is the fact that the Condorcet winner can't finish last in the aggregated pairwise vote. The complete explanation of this result is given in the Feature Box on page 117.

In case you were wondering why the subtitle of this section is "the Borda count strikes back!" we are ready to give the answer: The aggregated pairwise voting procedure is equivalent to the Borda count (Saari, 1994). The strengths of the aggregated pairwise vote are thus inherited by the Borda count. Saari advocates the use of the Borda count for a number of reasons, but the fact that it "is the 'natural' extension of the standard vote between two candidates" (Saari, 1994, p. 178) to an election with more candidates is one of the most persuasive reasons.

How do we know that the two procedures equivalent? Consider a voter i for whom $x \succ_i y \succ_i z$. The aggregation shows that i casts two votes for x (i prefers x to both y and z). Similarly, i votes for y against z, and the aggregation records one vote for y. This is, of course, exactly the same as casting a Borda ballot vector (2, 1, 0). The conclusion is that the Borda count has the same information that is collected by the aggregation of pairwise votes.

The fundamental difference between plurality/majority voting and the Borda count is laid bare by this discovery. Recall the definition of plurality rule, which is the same as majority rule if none of the voters are indifferent.

$$\text{Plurality rule}: x \succ_P y \text{ if and only if } |\{i \in N : x \succ_i y\}| > |\{i \in N : y \succ_i x\}| \qquad (3.6)$$

Note that the first terms on each side of the following inequality are the same:

$$\text{Borda rule}: x \succ_{BC} y \text{ if and only if}$$
$$|\{i \in N : x \succ_i y\}| + |\{i \in N : x \succ_i z\}| > |\{i \in N : y \succ_i x\}| + |\{i \in N : y \succ_i z\}| \qquad (3.7)$$

The Borda comparison between x and y always includes some information about the irrelevant alternative z. The violation of the independence from irrelevant alternatives that we encounter in examples involving the Borda count appears to be inherent within it. At the same time, however, the Borda count avoids both intransitivities and the danger of allowing a Condorcet loser to win an election.

Feature Box: The Aggregated Pairwise Vote and the Borda Count

In *Geometry of Voting*, mathematics professor Donald Saari argues that the Borda count is less prone to peculiarities than other voting methods that are based on ranked lists. The Borda count admits only 135 of 351 possible words (combinations of pairwise and listwise outcomes). In contrast, other methods of tabulation allow anything (literally!) to happen. A central part of his argument is the fact that the Borda count is equivalent to aggregated pairwise voting, and in the latter we have observed that a Condorcet winner never places last.

To see why only 135 of the possible words can occur, begin by considering all the Borda orderings with three alternatives. They are divided into three columns, representing social orderings in which the number of ties is zero, one, or two.

1. $x \succ_{BC} y \succ_{BC} z$	7. $x \succ_{BC} y \approx_{BC} z$	13. $x \approx_{BC} y \approx_{BC} z$
2. $x \succ_{BC} z \succ_{BC} y$	8. $x \approx_{BC} y \succ_{BC} z$	
3. $y \succ_{BC} x \succ_{BC} z$	9. $y \approx_{BC} z \succ_{BC} x$	
4. $y \succ_{BC} z \succ_{BC} x$	10. $y \succ_{BC} z \approx_{BC} x$	
5. $z \succ_{BC} x \succ_{BC} y$	11. $z \succ_{BC} x \approx_{BC} y$	
6. $z \succ_{BC} y \succ_{BC} x$	12. $z \approx_{BC} x \succ_{BC} y$	

(continues)

We now have to figure out which pairwise outcomes are consistent with each of these Borda outcomes. The problem is attacked by the timeless method of "divide and conquer."

Consider outcomes 1–6, the ones in which there are no ties. Keeping in mind the fact that a Condorcet winner cannot finish in last place in the Borda count (and the Condorcet loser cannot finish first), we can figure out which pairwise comparisons are possible. We focus on this one case $x \succ_{BC} y \succ_{BC} z$ (and will generalize to the others).

Possible if	$x \succ_{BC} y \succ_{BC} z$	Not possible if $x \succ_{BC} y \succ_{BC} z$
$x \succ_P y, y \succ_P z, x \succ_P z$	$x \approx_P y, y \succ_P z, x \succ_P z$	$x \approx_P y, y \succ_P z,$ $x \approx_P z$
$x \succ_P y, y \succ_P z, z \succ_P x$	$x \approx_P y, y \approx_P z, x \succ_P z$	$x \approx_P y, y \succ_P z,$ $z \succ_P x$
$x \succ_P y, y \succ_P z, z \approx_P x$	$x \approx_P y, z \succ_P y, x \succ_P z$	$x \approx_P y, y \approx_P z,$ $z \succ_P x$
$x \succ_P y, z \succ_P y, z \approx_P x$	$y \succ_P x, y \succ_P z, x \succ_P z$	$x \approx_P y, y \approx_P z,$ $z \approx_P x$
$x \succ_P y, y \approx_P z, z \succ_P x$	$y \succ_P x, y \approx_P z, x \succ_P z$	$x \approx_P y, z \succ_P y,$ $z \succ_P x$
$x \succ_P y, y \approx_P z, z \succ_P x$	$y \succ_P x, z \succ_P y, x \succ_P z$	$x \approx_P y, z \succ_P y,$ $z \approx_P x$
$x \succ_P y, y \approx_P z, z \approx_P x$	$y \succ_P x, z \succ_P y, z \succ_P x$	$y \succ_P x, y \succ_P z,$ $z \succ_P x$
$x \succ_P y, z \succ_P y, x \succ_P z$	$y \succ_P x, z \succ_P y, z \approx_P x$	$y \succ_P x, z \succ_P y,$ $z \succ_P x$
$x \succ_P y, z \succ_P y, z \succ_P x$		$y \succ_P x, z \approx_P y,$ $z \succ_P x$
		$y \succ_P x, y \succ_P z,$ $x \approx_P z$

There are 17 sets of pairwise contests that are consistent with $x \succ_{BC} y \succ_{BCM} z$. Because the setup of the problem is completely symmetric, the same must be true of Borda orderings 2–6. Because $6 \cdot 17 = 102$, we have found 102 of the 135 legal words.

Next consider the six Borda outcomes that have only one tie, which are items 7–12. There are 30 sets of pairwise outcomes that are consistent with these Borda results (5 for each one). Considering Borda ordering 7, we find that the following are the legal words.

$$
\begin{aligned}
&1.\ x \succ_P y, \quad x \succ_P z, \quad y \approx_M z, \quad x \succ_{BC} y \approx_{BC} z \\
&2.\ x \succ_P y, \quad x \succ_P z, \quad y \succ_P z, \quad x \succ_{BC} y \approx_{BC} z \\
&3.\ x \succ_P y, \quad z \succ_P x, \quad y \succ_P z, \quad x \succ_{BC} y \approx_{BC} z \\
&4.\ x \succ_P y, \quad x \approx_M z, \quad y \succ_P z, \quad x \succ_{BC} y \approx_{BC} z \\
&5.\ x \approx_M y \quad x \succ_P z, \quad y \succ_P z, \quad x \succ_{BC} y \approx_{BC} z
\end{aligned}
\tag{3.8}
$$

With five more legal words for each of the six Borda outcomes, we have thus added $5 \cdot 6 = 30$ valid words. The proof of this result is left as an exercise for the reader (but it is very informative, and so a complete answer is included at the end of the book).

Finally, consider Borda ordering 13. If the Borda count yields total social indifference, $x \approx_{BC} y \approx_{BC} z$, only three sets of pairwise matchings are possible:

$$
\begin{aligned}
&1.\ x \succ_P y, \quad y \succ_P z, \quad z \succ_P x \\
&2.\ y \succ_P x, \quad z \succ_P y, \quad x \succ_P z \\
&3.\ x \approx_P y, \quad y \approx_P z, \quad z \approx_P x
\end{aligned}
\tag{3.9}
$$

The first two are the cyclical pairwise comparisons, and the final is generalized indifference. With those three legal words, our total is now 135.

3.6.4 The Schulze Method

If you search the Internet, this one is often found under the unseemly titles "Clone-proof Schwartz Sequential Dropping" or the "Beatpath Method." We have named it for its originator, Markus Schulze, who has been an active proponent of the method since 1997 and has shown that it has many desirable qualities (Schulze, 2003). This method evolved in a sequence of e-mail and Internet postings by a group of enthusiasts who sought to develop workable voting methods that can actually be put to use in "real life." Schulze notes that this method is used in organizations that have an aggregate membership of more than 1700 and that it is the most widely used of all of the Condorcet round robin methods. People who use the Linux operating system might be interested to know that the Schulze method is used in decision making in the Debian Linux project, one of the most widely disseminated Linux distributions.

Of all the methods we have considered so far, this is the most difficult to understand and interpret at face value. To justify it, the proponents do not rely on intuition.

Rather, they rely on the mathematical fact that this procedure meets many of the criteria against which voting procedures are judged. If there is a Condorcet winner, this method selects it as the winner. The Schulze method satisfies many of the other most important priorities, including undifferentiatedness (same as neutrality), anonymity, monotonicity, independence of clones (cloneproofness), and several others.

Schulze's method also satisfies a property known as Smith-IIA. Recall the Smith criterion, which held that the winner should be a member of the Smith set. Smith-IIA adds the additional stipulation that if a candidate is added in an election, it can only affect the election outcome if it is in the Smith set. If it is not in the Smith set, then it is an irrelevant alternative (so far as the choice among alternatives in the Smith set is concerned) and it should not matter.

The voters submit rankings of as many candidates as they wish, and the ones they do not rank are assumed to be interchangeable and less desirable than the ones they do rank. There are two mathematically equivalent descriptions of the Schulze method for picking the winner in a competitive election. These methods use the idea of "transitive defeat." A "transitive defeat" (also known as a chain) allows x to assert itself against y through a chain of comparisons, like $x \succ_P w$ and $w \succ_P y$. Even if y defeats x in a head-to-head comparison, if a chain from x to y exists, then x can "transitively defeat" y. You might think of the chain as a sort of "self-defense" or "political countermeasure" for electoral candidates. The chains can be quite long, and to save space we might as well write $x \succ_P w \succ_P z \succ_P y$ to mean that x transitively defeats y with w and z in the middle of the chain.

As the old saying goes, a chain is only as strong as its weakest link. The strength of a chain from x to y is measured by the smallest margin of victory between any two alternatives in the chain. The symbol $P[x, y]$ represents the strength of the strongest chain from x to y. If there are two chains leading from x to y, we measure the strength of each one (i.e., find the smallest margin of victory), and $P[x, y]$ is the largest value. If there is no chain from x to y, then the strength is $P[x, y]=0$.

The "beatpath method" is the one that Schulze uses as the definition of the social ranking process. The aim is to find an alternative with the strongest chains through which it can defeat all other alternatives. If the chain from x to y is stronger than the chain from y to x, $P[x, y] > P[y, x]$, then x disqualifies y from further consideration. If we collect up all the alternatives that are not disqualified, then we have a set of "potential winners." If we are lucky enough to have only one candidate left, then it is the beatpath winner. If we have several left, then a tie-breaking procedure is needed. Quite frankly, the emphasis in this method is to narrow the alternatives down to this set of unbeaten candidates, and the tie-breaking procedure is something of an afterthought.

To bring this down to earth, consider the application of this method to choosing a national football champion after gathering votes from sports writers and coaches. The voters rank three undefeated teams, Auburn (AU), the University of Southern California (USC), and the University of Oklahoma (OU). In Figure 3.4, the arrow from USC to OU indicates that USC is preferred to OU by 44 of the voters. Here, we are using the number of voters as the indicator of the strength of the defeat (we

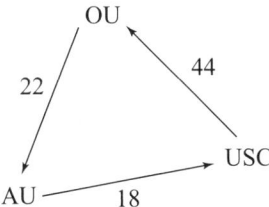

Figure 3.4 *Beatpath for college football.*

could use the margin instead). Note there is a cycle, where OU is ranked higher than AU (strength = 22) and AU is preferred to USC (strength = 18). This example is particularly simple because there is only one path from each alternative to each of the others. The beatpaths are as follows:

1. P [OU, AU] = 22
2. P [OU, USC] = 18
3. P [AU, USC] = 18
4. P [AU, OU] = 18
5. P [USC, OU] = 44
6. P [USC, AU] = 22

Because P [USC, OU] > P [OU, USC] and P [USC, AU] > P [AU, USC], USC is the beatpath winner.

The implications of the beatpath calculations are difficult to grasp for many people. Why does the procedure satisfy so many of the desirable qualities? How can we make sense of it? Shortly after Schulze proposed his method in 1997, it was shown that it is equivalent to a method in which we sequentially measure the strength of the candidates and then remove the weak ones, and then reevaluate. The sequential evaluation process is organized around an enhancement of the Smith set, which is called the Schwartz set. This approach is often called clone-proof Schwartz sequential dropping (CSSD) because we sequentially recalculate the Schwartz set after penalizing the weak candidates.

To be a member of the *Schwartz set*, an alternative must be either undefeated in pairwise competition (it may defeat *or tie* each other alternative) or able to transitively defeat every other alternative that can transitively defeat it. To be in the Schwartz set, a candidate must have "electoral self-defense" against each of the other candidates. More formally, x is in the Schwartz set if for any y such that P [y, x] > 0, then P [x, y] > 0. Even if x transitively defeats y, y can be in the Schwartz set if there is a chain of plurality decisions reaching from y to x. An alternative y is rejected from the Schwartz set if y cannot transitively defeat another alternative that can transitively defeat y. If there are no pairwise tie votes, the Schwartz set equals the Smith set. If there are ties, then the Schwartz set can be much smaller than the Smith set. In Figure 3.5, ties are represented by dotted lines that connect alternatives. The ties

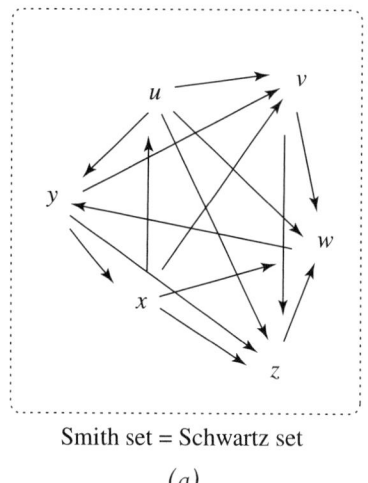

Smith set = Schwartz set

(a)

(b)

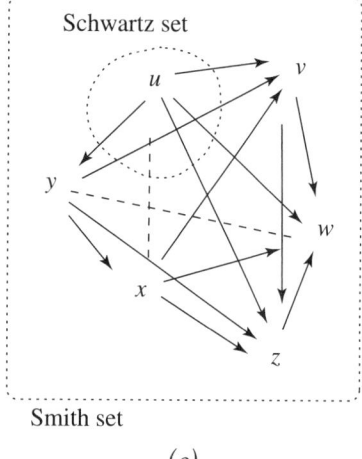

Figure 3.5   *Schwartz sequential dropping. (a) Smith set = Schwartz set. (b) One tie shrinks the Schwartz set. (c) Schwartz set shrinks even further.*

(c)

make a difference because the transitive defeat concept is based on pairwise victories and ties are treated the same as losses. In other words, ties break chains.

The sequential approach can be thought of as a way to find the strongest chains. The Schwartz sequential dropping version of the Schulze procedure is the one that is described in the Constitution of the Debian software organization *http://www.debian .org/vote/2003/vote_0002*. In Figure 3.5, we illustrate the sequential narrowing of the Schwartz set that results when ties (dotted lines) are inserted.

Schwartz sequential dropping proceeds iteratively, repeatedly narrowing down the alternatives. It is easy to make this sound more complicated than it really is, but we will try not to.

Step 1. Collect the ranking information from the voters. Voters are allowed to rank all the alternatives or just some of them. They are allowed to register their indifference between alternatives. From the ballots, we can calculate the pairwise outcomes for all alternatives, either $x \succ_P y$ or $x \approx_P y$ or $y \succ_P x$.

Now the decision-making starts.

Step 2. Eliminate rejects. Begin by simplifying the problem by isolating "rejects," alternatives that are losers in the round robin tournament. These losers are the ones left out of the Schwartz set. Recall that y is rejected if it is transitively defeated by some other proposal and there is no chain that leads from y to that other alternative. The rejects are eliminated from consideration, permanently.

Step 3. Evaluate the remaining contenders (members of the Schwartz set).

A. If there is only one remaining contender, then it is the winner.

B. If there is more than one contender, check the pairwise comparisons among all the contenders. If all pairs are tied, it means, in some sense, that we have found all the alternatives that are "equally good." If all are tied, proceed to the tie-breaking procedure in step 4.

C. If all contenders are not tied pairwise, then we need to make some changes so we can reject more alternatives. Here is where the "sequential dropping" kicks in. Among all pairs of contenders, find the "weakest pairwise defeat." (To us, a name like "least impressive victory" would be more fitting.) The idea is to spot the least worthy pairwise victor and demote it. How to spot the least impressive winner? Look for the alternative that wins with the smallest number of votes.[3] When the weakest pairwise defeat is found, that victory can be erased by changing it to a tie in the data. That is, we change the result in the round robin data for the weakest defeat from $\succ_P$ to $\approx_P$. (Caution: Contrary to what some websites say, neither the weakest defeat nor the tied alternatives are not dropped literally. The weakest defeat is converted to a tie and the Schwartz set is recalculated.) If there are several alternatives that are equally weak winners—many weakest defeats exist—then choose the one with the most votes for the loser and declare that one to be the weakest defeat. If there are several pairs tied in both the number of votes for the winner and the loser, treat them all in the same way: Change the pairwise comparison to a tie. This generates new round robin data.

Go back to step 2 and repeat the procedure. Use the new round robin data, of course. The rejects that were eliminated in previous stages are never readmitted. Because we have inserted some ties into the results, the members of the Schwartz set

---

[3]Some versions of this procedure use the margin of victory rather than the absolute number of votes received by the winner. Schulze recommends the total number of votes because he believes it gives voters a positive incentive to report their rankings as completely as possible. Otherwise, they might have an incentive to report just the rankings of the top one or two alternatives to "game" the election system.

will change: Some contenders will become rejects and then they are dropped. The procedure will repeatedly loop between steps 2 and 3 until the condition described in step 3B is obtained.

Step 4. If we arrived at this stage, it means we have narrowed down the alternatives to the most appealing set of more or less equally good outcomes. Schulze calls these "potential winners." A winner is selected by a tie-breaking rule.

The tie-breaking rule that is suggested by Schulze is, more or less, going to rank the tied outcomes by listening to the opinions of randomly drawn voters. Suppose the tied candidates are $\{w, x, y, z\}$. Randomly draw a ballot and copy down the preferences of the voter. If that ballot states a ranking for all the alternatives, then that is the end of the procedure. The group result will match that ballot. If the ballot is only partially filled in, for example, it states only $x \succ_i y$, then we take that information and mark down $x \succ_{\text{Schulze}} y$ (it is "set in stone," unalterable) and we draw another ballot. If that ballot has preference information about candidates that are not already sorted by $\succ_{\text{Schulze}}$, such as w and y, then use that information $w \succ_{\text{Schulze}} y$. We draw ballots at random until every pairwise comparison is filled in, and we *never* change any of the Schulze comparisons we have already completed. If we happen to go through all ballots without ranking all the alternatives, then the final ranking is assigned at random.

The Schulze method can be applied to any of the voting examples we have considered. Of course, if the preferences imply the existence of a Condorcet winner, then that alternative is chosen and the solution is simple. To find a more interesting case, we need to select an example in which there is a voting cycle. Recall the Borda paradox considered in section 3.4.2. In Figures 3.6 and 3.7, we have used the preferences from Tables 3.3 and 3.4 with the Schwartz sequential dropping approach.

In Figure 3.6, we note that the original preferences allow a majority rule cycle across all of the alternatives, so the first-stage Schwartz set includes all the alternatives. There are two comparisons tied for the weakest defeat, each with the support of four voters, so in Figure 3.6b, those arrows are changed to dotted lines representing ties. After those ties are taken into account, the Schwartz set is recalculated, and it includes only w.

Recall that alternative z is exposed as a fraud in that example. The effect of dropping z to the bottom of the voter preference orderings appears in Figure 3.7. Note that the initial Schwartz set now includes three alternatives, $\{w, x, y\}$. Alternative z is excluded (it is a Condorcet loser). The weakest defeat is between w and y. After that is changed to a tie, then w is the only potential winner remaining. The same alternative, w, wins with the Schulze method in both cases.

The Schulze method has the apparent strength of solving the paradox in the Borda count: w wins in both cases. The Schulze method is thus more stable and, in some sense, more logical. The fact that the Borda method chose y in the original example (Table 3.3) and Schulze chooses w deserves mention. From a beatpath point of view, it appears that we did not correctly understand the Borda problem in section 3.4.2. We thought the paradox was that y was turned from a winner into a loser when the fortunes of z changed. Now it appears that the original Borda winner, y, was

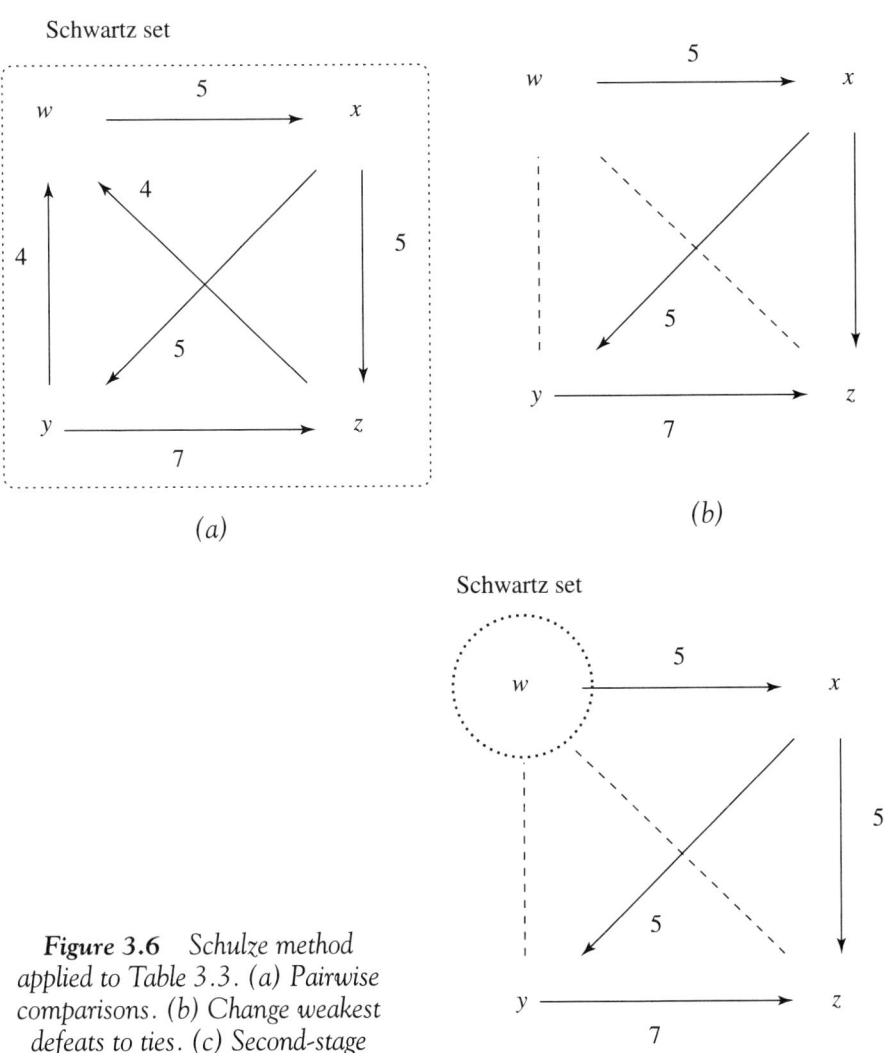

Figure 3.6 *Schulze method applied to Table 3.3. (a) Pairwise comparisons. (b) Change weakest defeats to ties. (c) Second-stage Schwartz set includes only w.*

"wrong." Borda chose the wrong winner because it allowed z to play the role of a spoiler, somehow distracting voters from their purpose, thus allowing y to win. When z is moved to the bottom in Table 3.4, its role (its influence as an irrelevant alternative) is eliminated and the "correct alternative" w wins.

The Schulze method leads to a full transitive ordering of all the candidates. The beatpath values $P[x, y]$ can be used to sort alternatives from top to bottom. As a result, the method could be used to determine the winner of the Division 1A College football championship on the basis of the rankings submitted by coaches, writers, or computers.

The reader might be wondering, "if this method is so great, why isn't everyone using it?" If you ask its proponents, we expect they will say "because everyone is not

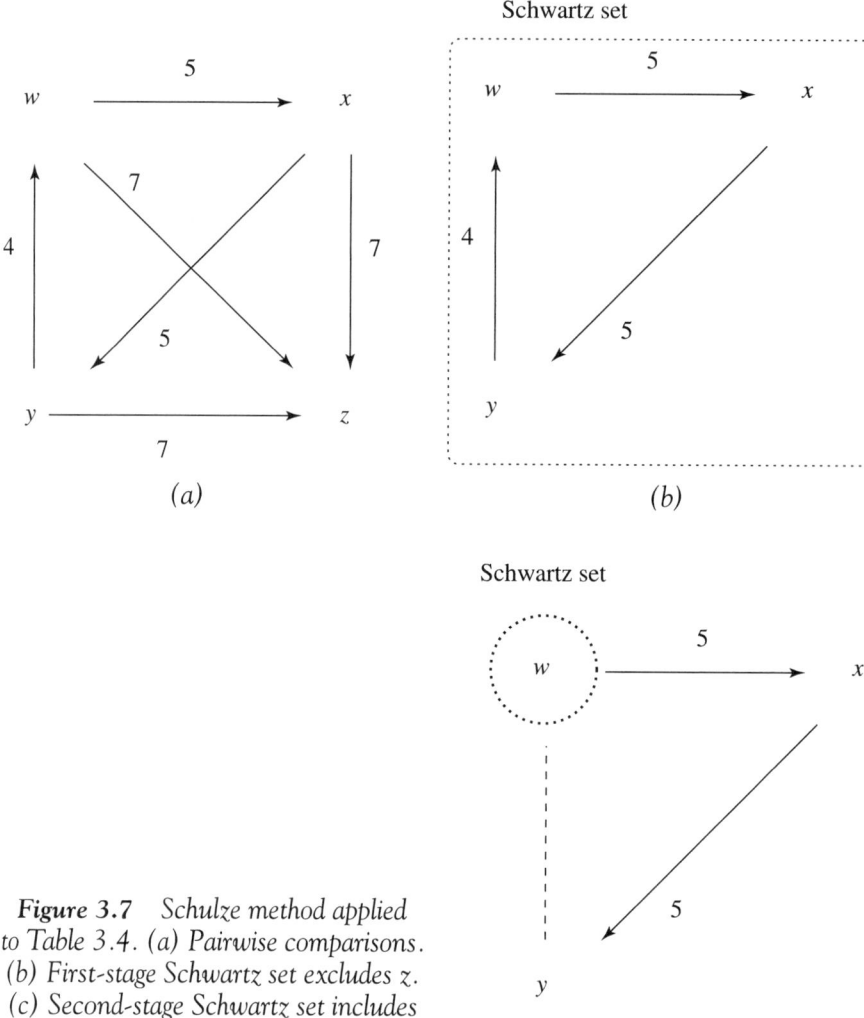

Figure 3.7 *Schulze method applied to Table 3.4. (a) Pairwise comparisons. (b) First-stage Schwartz set excludes z. (c) Second-stage Schwartz set includes only w.*

very smart!" Ordinary people—not academics—are not particularly interested in election procedures, much less complicated ones. As a result, the most widely used argument against the method seems to be that it is too complicated. To academics, this is not a very good argument, but it carries a lot of weight with real-life policy-makers. In response, the Schulze proponents will point out that, from a voter's point of view, it is no more complicated than a Borda count. All the voters need to do is submit rankings of some alternatives. The counting process is more complex, of course, but the effort is justified. The proponents can list a long list of favorable qualities that we have already mentioned, plus some more.

The beatpath approach is not completely free from the anomalies that infect the other voting mechanisms. Schulze draws our attention to the fact that his method violates the *participation criterion*. The participation criterion says that if an election has been decided and then you add voters who prefer the winner to another alternative, then adding those new voters must not result in the victory of that other (less favored) alternative. More formally, the winner is x, hence socially $x \succ_{\text{Schulze}} y$. Add more voters such that $x \succ_i y$. Adding those voters should not change the social result to $y \succ_{\text{Schulze}} x$. The criterion does not require that x must still win, only that adding voters who do not prefer y should not make y the winner. Please note that the participation criterion is different from the monotonicity principle. Monotonicity states that *among existing voters*, a change in voter preferences that favors x against y must not hurt x's position in the final decision. Participation concerns the addition of new voters. Schulze is a monotonic procedure, but examples can be found to show that it violates the participation criterion.

Schulze's example of the violation of the participation criterion is presented in an example that has six candidates, $\{u, v, w, x, y, z\}$ (Table 3.11). The winner is u when the electorate includes the 60 voters from groups 1–12. The violation of the participation principle is shown when group 13, which has three members, is added to the electorate. For all the members of group 13, note that u is the most preferred alternative and x is the least desirable. One would expect that their favorite, the winner u, would remain in good position. Unfortunately, the Schulze rule applied to this 63 voter electorate declares x to be the winner. After the sequential dropping process, u is the only contender (the only member of the final Schwartz set), and no use of the "tie breaker" is required.

Among the collection of peculiar election outcomes that we have been accumulating, this is surely one of the most bizarre. To investigate it more carefully, consider the graphs that illustrate the sequential dropping process in Figure 3.8. On the

Group	1	2	3	4	5	6	7	8	9	10	11	12	13
Number of voters	4	2	4	2	2	2	4	12	8	10	6	4	3
Preferences													
Favorite	u	u	u	u	v	w	w	x	y	z	z	z	u
Second	v	v	y	y	z	x	x	y	w	u	u	y	y
Third	w	z	v	z	u	v	v	w	x	v	v	x	z
Fourth	x	x	z	v	w	y	z	u	v	w	x	v	w
Fifth	y	y	w	w	x	z	y	v	z	x	y	w	v
Sixth	z	w	x	x	y	u	u	z	u	y	w	u	x

Table 3.11   The Schulze method

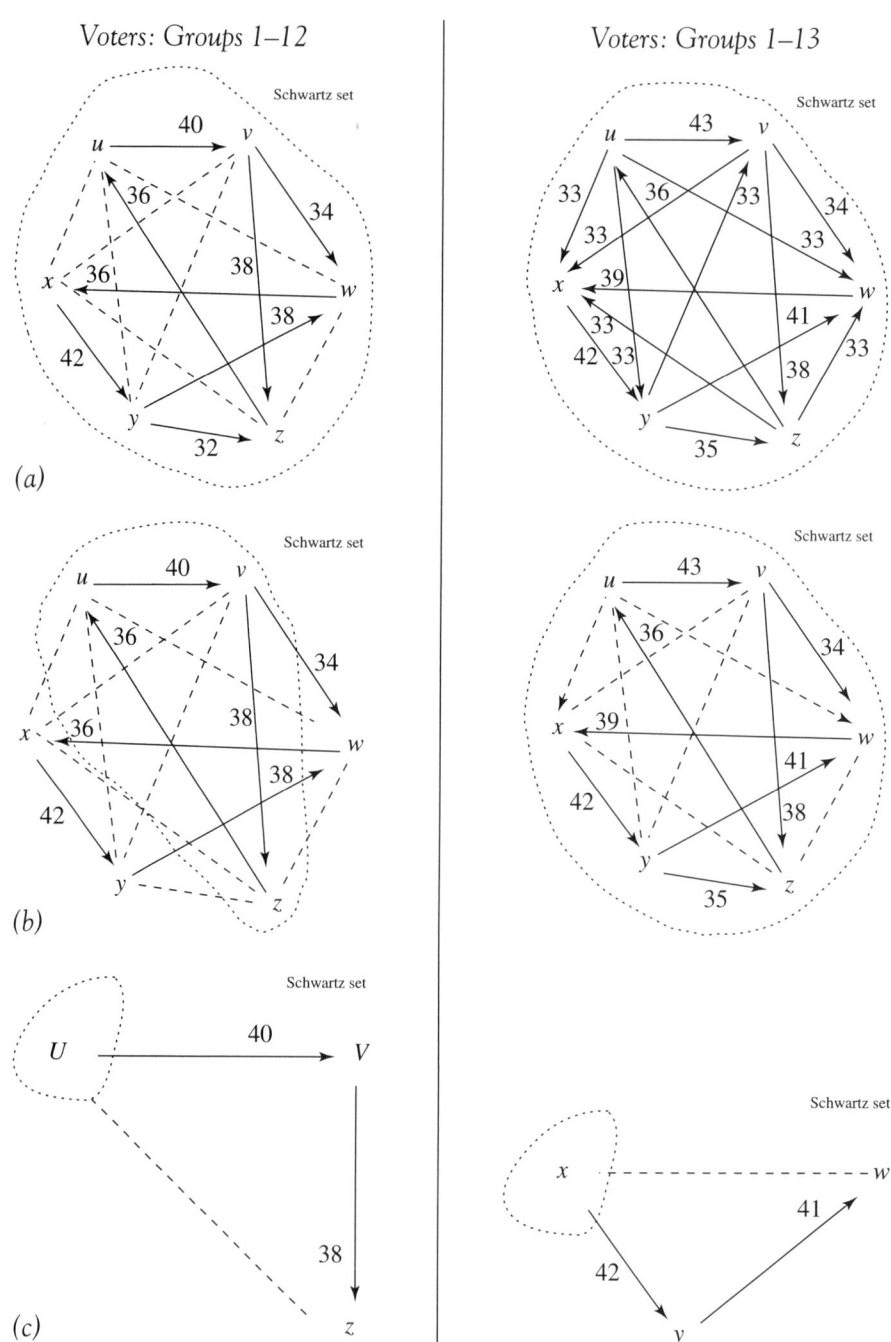

Figure 3.8 *Schulze method with preferences in Table 3.11. (a) Pairwise comparisons. (b) Change the weakest defeats to ties. (c) Final stage.*

left, the pairwise data for the 12-group problem is illustrated. Note there are many ties in this example. The weakest defeat is the step $y \succ_p z$ which is supported by 32 voters. After that is changed to a tie, the Schwartz set shrinks to $\{u, v, w\}$, and from there it is easy to drop the weakest defeat and see that u wins. On the right side of the figure, note that when group 13 is added, there are no pairwise ties. When the weakest defeats are changed to ties, then the pattern of ties and pairwise comparisons in the initial results on with 12 groups reappears. The ties are the same, but the margins of victory are different. With 13 groups, the weakest defeat is now $v \succ_p w$. After a tie is drawn between v and w, the Schwartz set shrinks to $\{x, w, y\}$, and from there the result is easy to see: x wins. Why did u lose? When the path from v to w was removed, then u lost its chains of self-defense. The alternatives x, y, and z can all transitively defeat u, and there is no path leading from u back to them if $v \approx_p w$.

This is a variant of the problem of independence from irrelevant alternatives. What is driving the problem here is not the relative evaluation of u and x but rather the careful placement of u, v, w, and z in the orderings of group 13 voters. The careful placement ends up changing the magnitude of pairwise victories, thus impacting the order in which the sequential dropping eliminates candidates. If you think of the sequential dropping process as a series of games in a tournament, then, we have simply found another example of the way in which the ordering of the matches affects the outcome.

EXERCISES 3.6

1. The following table describes the tastes of three voters. Conduct a round robin tournament and use the win–loss records to try to decide which candidate should win the tournament.

	Voters		
	1	2	3
Ranking			
First	x	y	z
Second	y	z	x
Third	z	x	y

2. Consider the table in the previous question. Describe a couple of changes you might make in voter preferences (either by changing voters or adding new ones) if you want z to become the winner in the round robin.

3. For the following preference profile, conduct a round robin tournament and calculate the aggregated pairwise vote totals for the candidates. Confirm that the results are the same as the Borda count.

	Voters		
Groups	1	2	3
Number of members	22	12	31
Ranking			
First	x	y	z
Second	y	x	y
Third	z	z	x

4. Consider the preferences used in the previous problem. Create a graph of the type shown in Figure 3.4. Use that graph to figure out which candidate would win with the Schulze procedure.

5. Consider the following preference profile. Create a beatpath graph of the type shown in Figure 3.4.

	Voters					
Groups	1	2	3	4	5	6
Number of members	200	150	150	300	100	100
Ranking						
First	z	y	x	w	y	x
Second	x	z	y	z	x	y
Third	y	x	z	x	w	z
Fourth	w	w	w	y	z	w

6. Use the graph you created in the previous question to figure out which alternatives are in the Schwartz set. If the Schwartz set contains more than one element, proceed with the sequential dropping process (change the weakest defeats to ties and recalculate the Schwartz set).

7. This is a challenging problem, one which we refer to in the feature box in Section 3.6.3. Show that if the Borda count (or, equivalently, the aggregated pairwise vote) outcome is $x \succ_{AP} y \approx_{AP} z$, then the only pairwise outcomes that are consistent with that outcome are the ones listed in equation (3.8).

3.7 Single Vote Systems: Cousins of Plurality and Majority Rule

We began this chapter with the plurality problem. A plurality election with 20 or 30 candidates might award victory to a very unpopular candidate. The poster children for the problem might be the Democratic and Republican candidates in Minnesota's gubernatorial election who were edged out by Jesse Ventura. In the words of mathematician Keith Devlin, Jesse Ventura "won not because the majority of the voters chose him, but because plurality voting effectively thwarted the will of the people. Had the voters been able to vote in such a way that, if their preferred candidate were not going to win, their preference between the remaining two would be counted, the outcome would have been quite different" (2004). The alternative schemes, the Borda rank order voting plan and the pairwise Condorcet systems, offer some hope. They have not won the hearts and minds of electoral reformers with the same power as the alternative to which we now turn our attention.

This system requires the voters to come to the polls only once. The voters rank the alternatives and then the election officials can proceed "as if" there really were a run-off election. When this procedure is used to select a single winner in an election, it is generally referred to as an instant runoff vote or an alternative vote system. When the method is used to select a group of candidates, such as the top five nominees for the Grammy Awards or in a proportional legislative election, the procedure is called the single transferable vote (STV). The procedures are the same for choosing a single winner, but the details get complicated for selecting the additional winners. There are two prominent implementations of the STV concept, the Hare system (in honor of Thomas Hare, an English proponent of this system) and the Hare-Clark system (add an honorific for Andrew Inglis Clark, a Tasmanian Attorney-General who pushed a variant of the STV into usage). The transferable vote is in use in national elections in Australia, Malta, and Ireland as well as in local elections in Cambridge, Massachusetts, San Francisco, California, cities in Northern Ireland, and New Zealand. The method is used in faculty elections at Cornell University.

We focus our attention on the selection of a single winner from a field of candidates, so we detail only that part of the procedure. The winner must obtain a target number of votes, dubbed the quota. If there is only one spot open—only one candidate can win—the usual quota is the majority of the votes:

$$\text{quota} = \text{majority} \geq \frac{\text{number of voters} + 1}{2}$$

On election day, the voters go to the polls and rank the candidates from first to last in order of desirability. Then election managers name the winner by applying the following algorithm.

Step 1. A voter's support on the first ballot is given to the voter's top-ranked candidate. If one of the candidates has a majority, then that candidate is the winner. If no candidate has a majority, proceed to step 2.

Step 2. The candidate with the smallest number of votes is eliminated from consideration. The ballots of the voters who had supported that eliminated candidate are reconsidered. Their votes are transferred to their second-ranked candidates. Then the votes are retabulated. If a candidate has a majority, then it is declared the winner. Otherwise, repeat step 2. (The candidate that was eliminated in this step is excluded throughout the remainder of the process.)

If the election is choosing five members of a commission or a legislative delegation, the procedure is slightly more involved. Not only are the votes cast for the losers transferred, but also the "excess" votes (ones which exceed the quota) received by the winners can be transferred.[4] The calculations in that reallocation are a bit involved; that is where the variants of STV diverge. The main idea in the procedure is that candidates are removed one by one, and the votes cast for the removed candidates are either counted for them (as part of their quota) or transferred and counted for other candidates that the voter also supports.

The proponents of this method claim that it does not throw away information in the same way as a plurality election. It saves voters from the terrible choice that is forced on them by a plurality vote. If the voters who favor either the Democrat or the Republican against Jesse Ventura have a chance to declare that priority, then Ventura might not end up winning in the instant runoff. The "Ventura problem," coupled with the ongoing use of the procedure in some places, gives the STV advocates some persuasive ammunition. Krist Novoselic, most well known as a member of the band Nirvana, has written a book that combines his band memoir with an endorsement of transferable votes, *Of Grunge and Government: Let's Fix This Broken Democracy* (2004). He also hosts a website (*http://www.fixour.us*). He eagerly watched the introduction of STV in San Francisco for the 2004 Board of Supervisors elections. Steven Hill, of the Center for Voting and Democracy, a proponent of the transferable vote systems, enthusiastically proclaimed, "More voters picked their supervisor, and fast results mean that San Francisco avoided a December runoff election, saving millions in taxes (the cost of the second election)" (Novaselic, 2004).

There is no doubt that the STV might work to correct bad outcomes in some particular examples. Perhaps Jesse Ventura might not have been governor of Minnesota. There are example problems that we can consider in which the STV does not seem to perform too badly. If the STV procedure is applied to the preferences of groups 1–12 in Table 3.11 (the data for the Schulze example that violated the participation criterion), the STV stages eliminate v, then w, then y, and then u. The two remaining candidates, x and z, are tied with 30 votes. After adding group 13 into the calculations, the elimination process rejects the alternatives in the same order (v, w, y, and then u). The winner is z, which earns 33 votes against the 30 for x. The new voters prefer z to x, and it seems quite reasonable that adding them to the electorate would tip the scales in favor of z.

On the other side of the ledger, however, the STV has several properties that should cause concern. The fact that the STV system does not always select a Condorcet winner (when one exists) is a serious shortcoming. Academic research has often concentrated on another issue: STV is not monotonic. The principle of monotonicity has first come to our attention in the consideration of majority rule and May's

---

[4]
$$\text{quota} = \frac{\text{number of votes cast}}{\text{number of winners to be chosen} + 1} + 1$$

Note that if there is one winner to be chosen, the winner must have a bare majority.

	Voters					Alternative preferences
Groups	1	2	3	4	5	3'
Number of members	5	4	2	6	9	
Ranking						
First	z	y	y	x	w	x
Second	x	z	x	y	z	y
Third	y	x	z	z	x	z
Fourth	w	w	w	w	y	w

Table 3.12   Single transferable vote violates monotonicity

theorem. Now it comes to the forefront again. The basic idea of the monotonicity criterion is that an election system should be responsive, and not in a perverse way. If a voter changes his or her mind and decides to rank an alternative more highly, then that alternative's electoral fortunes should be helped, or at least not hurt. This is the property commonly called weak monotonicity.

Consider the preference rankings in Table 3.12. There are preferences for five groups, {1, 2, 3, 4, 5}, and four alternatives, {w, x, y, z}. The STV algorithm proceeds as follows:

1. The first place votes are, in order, $w = 9$, $x = 6$, $y = 6$, $z = 5$.

2. Eliminate z, so the candidates are w, x, and y. Reallocate the votes of group 1 to their second-ranked candidate, x. The votes are retabulated: $x = 11$, $w = 9$, $y = 6$.

3. Eliminate y, so the final two candidates are w and x. Reallocate the votes of groups 2 and 3 to x. The final vote totals are $x = 17$, $w = 9$. So x is declared the winner.

We make only one small change to demonstrate the fact that STV is nonmonotonic. We replace group 3 with the voters in group 3', so that the columns being used in the vote are {1, 2, 3', 4, 5}. We hope the reader will agree that this seems to be an innocuous change. The voters in group 3 believe x is second best, but the voters in group 3' believe that x is the most desirable of all; x should still win, shouldn't it?

Observe how the STV result changes.

1. The first place votes are, in order, $w = 9$, $x = 8$, $y = 4$, $z = 5$.

2. Eliminate y, so the candidates are w, x, and z. Reallocate the votes of group 2 to their second-ranked candidate, z. The votes are thus retabulated: $z = 9$, $w = 9$, $x = 8$.

3. Eliminate x, so the final two candidates are z and w. Reallocate the votes of groups 3′ and 4 from x to z. The final vote totals are $z = 17$, $w = 9$. The winner is z.

This is a clear violation of weak monotonicity (and, we might add, plain old common sense). The winner of the original STV was x, but after we raise x's popularity by replacing 3 with 3′, the winner is z. Alternative x was eliminated in stage 3. This would be interpreted to mean that x finished in third place. And the only change between the first and second example was that the voters in group 3 became group 3′, meaning they raised x in their preferences. After noting this fact, William Riker concluded: "It is hard to believe that there is any good justification for the single transferable vote" (1982, p. 51).

EXERCISES 3.7

1. Consider the preferences used in problem 2 of Section 3.4. Conduct an STV election.

    a) In what order will the candidates be eliminated? How will the votes be reallocated?

    b) Who will the eventual winner be?

    c) If you conduct an ordinary runoff election with the top two vote getters, does the result agree with that of the instant runoff?

2. Repeat the previous exercise with the preferences used in problem 3 of Section 3.4.

3. Repeat the previous exercise with the preferences used in problem 2 of Section 3.5.

4. The previous problems should lead you to expect that the instant runoff gives the same result as the ordinary majority/runoff election procedure. With four candidates, however, the same is not true. Conduct elections with these voter preferences to see an example.

		Voters					
Groups		1	2	3	4	5	6
Number of members		100	250	150	300	100	100
Ranking							
First		z	y	x	w	y	x
Second		x	z	y	z	x	y
Third		y	x	z	x	w	z
Fourth		w	w	w	y	z	w

5. Create a voter preference profile in which three candidates are considered and there is a Condorcet winner. Then calculate the instant runoff election winner. Experiment with your example until you find a profile in which the Condorcet winner does not win the instant runoff.

6. Suppose you are a member of the Book-of-the-Month club. The club must choose between the classics *The Grapes of Wrath* and *Superman: Battle to the Finish with Luther!* You would love to have *Superman*, but you know for sure that at least 70% of the club members are snobs who would vote for *Grapes*. Think of two methods you might try in order to bring about the eventual selection of *Superman*. Note that you are not allowed to make a persuasive speech to change their minds or do anything that would change the pairwise comparison of the two books, but you can introduce new alternatives and different voting methods.

3.8 Conclusion

After reviewing the many voting systems and studying the many principles by which they are judged, the reader has a right to ask the author, "How do you rank these systems? Which is the best?"

We find ourselves wishing that there were some voting system that would meet all the requirements that seem fair and logical. Wishing does not make it so, sadly enough. In fact, there is a theorem, known as Arrow's possibility theorem, which states, basically, that it is mathematically impossible to find a voting system that satisfies what appear to be the most basic bare-minimum requirements. Arrow, an economist, was awarded the Nobel prize, partly on the basis of this research (Arrow, 1963).

Arrow's theorem can be stated in several ways. We believe this is the most useful version. Suppose there is an election ranking system that meets these requirements:

Nondictatorial: The social ranking does not mirror the tastes of one voter i without regard to the tastes of the other voters.

Pareto Principle: If all voters favor one alternative over another, then the social ranking should respect that and rank the alternatives accordingly.

These seem like such simple fundamental elements that they can be assumed as necessary ingredients in a social ranking procedure. Arrow's possibility theorem says the following: If an election system is nondictatorial and obeys the Pareto principle, then it is possible to find an example set of voter preferences such that

1. The social ranking is intransitive (a voting cycle), and/or

2. Independence of irrelevant alternatives (IIA) is violated.

It is very important to emphasize that the possibility theorem does not mean that a ranking procedure is always intransitive or in violation of IIA. Rather, it is always possible to find an example society, a set of voter preferences, such that the result is somehow illogical. If the result is intransitive, then we have the paradox of "rational man, irrational society" (Barry and Hardin, 1982). If we adopt a system that blocks intransitivity, such as the Borda count, then we know for sure there are examples in which independence from irrelevant alternatives is violated.

Arrow's theorem is sometimes overstated. For example, some students believe it says that democracy is always illogical. It does not mean anything of the kind. Rather, it means that it is not possible to find a democratic procedure that works in a sensible way all of the time.

If it is not possible to meet such basic simple criteria as nondictatorship, of course it is not possible to meet even more elaborate properties like the participation criterion. As a result, the choice of electoral system has to be based not on which system meets all the specified criteria, but rather on a value judgment about what sorts of peculiarities are the most harmful and which sorts of societal preference profiles are most likely to arise and cause peculiarities.

If the reader wants to know which ranking system is best, we can sympathize. We want to know the same thing—and so do most political scientists, economists, mathematicians, physicists, and psychologists.

Personally, we have found pleasant results by using procedures that eliminate the obviously bad candidates (by some principle like the Smith set or the Schwartz set) and then choosing among the remainder by a positional method like the Borda count or the Schulze method. We have been very unhappy with decisions reached by pure plurality or a runoff election. The STV might be preferable to a plurality or majority rule election, but the fact that it is so badly nonmonotonic causes us to shy away from it. That is to say, plurality rule is truly bad, as both Borda and Condorcet had contended, and we would rather have STV than plurality, but we'd rather have one of the more elaborate two-stage procedures instead of either one.

Readers who look into this further will no doubt find 10 or 20 seriously proposed election procedures, all of which have advantages worth considering. Perhaps one would begin with the textbook by William Riker (1982) and then do some soaking and poking on the Internet. We have found many fascinating websites and would encourage readers to adventure; but keep in mind that these sites are not peer reviewed by experts, and we have noticed a nontrivial number of mistakes in interpretation and mathematics. Perhaps the most rigorous analysis is to be found on the election methods e-mail list, which is hosted on *http://electorama.com*.

Chapter 4
Game Theory

The mathematical theory of games was first developed as a model for situations of conflict. It gained widespread recognition in the early 1940s when it was applied to the theoretical study of economics by the mathematician John von Neumann (1903–1957) and the economist Oskar Morgenstern (1902–1976) in their book, *Theory of Games and Economic Behavior*. Since then, its scope has been broadened to include cooperative interactions as well, and it has been applied to the theoretical aspects of many of the social sciences. Although the jury is still out on the question of whether this theory furnishes any valuable information regarding practical situations, it has stimulated much basic research in disciplines such as economics, political science, and psychology.

4.1   Introduction to Games

Situations of conflict, or any other kind of interactions, are called *games* and they have, by definition, participants who are called *players*. We limit our attention to scenarios where there are only two players, which we name Ruth and Charlie. The existence of a conflict is usually due to both the desire of each player to improve his or her circumstances, frequently by means of some acquisition, *and* the unfortunately limited nature of all resources. It is assumed that each player is striving to *gain* as much as possible and, for the next two sections, that each player's gain is his or her opponent's loss. Finally, each player is assumed to have several *options* or *strategies* that he or she can exercise (one at a time) as his or her attempt to claim a portion of the resources. Because of the introductory nature of this text, most of the foregoing discussion is restricted to situations wherein the players make their moves simultaneously and independently of each other.

The foregoing discussion is admittedly vague, and this section's remainder is devoted to the informal exposition of several examples of increasing complexity. The requisite formal definitions are postponed to the next section.

PENNY MATCHING.   *Ruth and Charlie each hold a penny and they display them simultaneously. If the pennies match in the sense that both show* heads *or both show* tails, *then Ruth collects both coins. Otherwise, Charlie gets them.*

It is clear that this game is fair in the sense that neither player has an advantage over the other. Moreover, if the game is only played once, then neither player possesses a shrewd system, or strategy, that improves his or her position, nor is there a foolish decision that worsens it.

The situation changes if the game is played many times. Although the repeated game remains symmetric, and neither player has a strategy that guarantees his or her coming out ahead in the long run, it *is* possible to play this game foolishly. Such would be the case were Ruth to consistently display the *head* on her penny. In that case Charlie would be sure to catch on and display the *tail* each time, thus coming out ahead. It is a matter of common sense that neither player should make any predictable decisions, nor should a player favor either of his or her options. In other words, when this game is repeated many times each player should play each of his or her options with equal frequency 1/2 and make his or her decisions unpredictable. One way a player can accomplish this is by flipping his or her coin rather than consciously deciding which side to show. With an eye to the analysis of more complex games, this game is summarized as follows:

Charlie

		Heads	Tails
	Heads	1	−1
Ruth			
	Tails	−1	1

The entry 1 denotes a gain of one penny for Ruth and the entry −1 denotes a loss of one penny for Ruth. Because Ruth's gain is Charlie's loss, and vice versa, this array completely describes the various outcomes of a single play of the game.

All the subsequent examples and most of the discussion is phrased in terms of the same two players, Ruth and Charlie. Payoffs are described from Ruth's point of view. Thus, a payoff of a penny always means a penny gained by Ruth and lost by Charlie.

ROCK-SCISSORS-PAPER.   *Ruth and Charlie face each other and simultaneously display their hands in one of the following three shapes: a fist denoting a* rock, *the forefinger and middle finger extended and spread to suggest* scissors, *or a downward facing palm denoting a sheet of* paper. *The rock wins over the scissors because it can shatter them, the scissors win over the paper because they can cut it, and the paper wins over the rock because it can be wrapped around the latter. The winner collects a penny from the opponent and no money changes hands in the case of a tie.*

This game has features that are very similar to those of penny matching. There are neither shrewd nor foolish decisions for a single play. If the game is repeated many times, then players who favor one of the options place themselves at a disadvantage. The best strategy for each player is to play each of the options with the same fre-

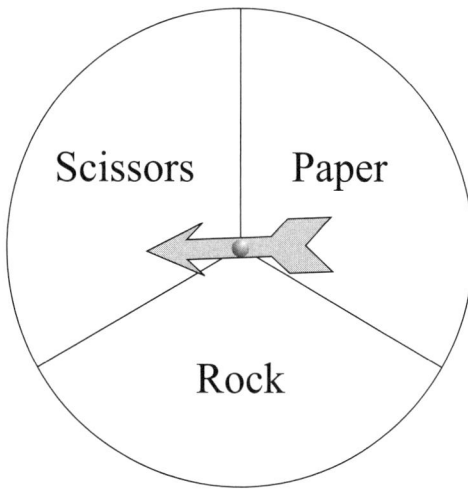

Figure 4.1   *A Spinner*

quency of 1/3 in a manner that yields the opponent as little information as possible about any particular decision. For example, he or she could base each decision on the result of a spin of the spinner of Figure 4.1.

The outcomes of the rock-scissors-paper game are tabulated as follows:

		Charlie		
		Rock	Scissors	Paper
	Rock	0	1	−1
Ruth	Scissors	−1	0	1
	Paper	1	−1	0

As before, a positive entry denotes a gain for Ruth, whereas a negative entry is a gain for Charlie.

The biologists B. Sinervo and C. M. Lively recently reported on a lizard species whose males are divided into three classes according to their mating behavior. The interrelationship of these three alternative behaviors very much resembles the rock-scissors-paper game and merits a digression here. Each male of the side-blotched lizards (*Uta stansburiana*) exhibits one of three (genetically transmitted) mating behaviors:

1. *Highly aggressive* with a large territory that includes several females;

2. *Aggressive* with a smaller territory that holds one female;

3. Nonaggressive *sneaker* with no territory who copulates with the others' females.

The highly aggressive male is stronger than the monogamous one who in turn is stronger than the sneaker. However, because the highly aggressive males must split their time between their various consorts, they are vulnerable to the sneakers. The observed consequence of this is that the male populations cycle from a high frequency of aggressives to a high frequency of highly aggressives, then on to a high frequency of sneakers and back to a high frequency of aggressives.

The next game is played in Italy using three fingers. For pedagogical reasons a two-finger simplification is examined here.

TWO-FINGER MORRA.   *At each play Ruth and Charlie simultaneously extend either one or two fingers and call out a number. The player whose call equals the total number of the extended fingers wins that many pennies from the opponent. In the event that neither players' call matches the total, no money changes hands.*

It clearly would be foolish for a player to call a number that cannot possibly match the total number of displayed fingers. Thus, a player who extends only one finger would call either 2 or 3, whereas a player who extends two fingers would call only 3 or 4. Consequently, each player has in reality only four options and the game's possible outcomes are summarized as follows:

Charlie

		(1, 2)	(1, 3)	(2, 3)	(2, 4)
	(1, 2)	0	2	−3	0
	(1, 3)	−2	0	0	3
Ruth	(2, 3)	3	0	0	−4
	(2, 4)	0	−3	4	0

where the option (i, j) denotes the extension of i fingers and a call of j.

This game, like its predecessors, is symmetric. Neither player has a built-in advantage. It is tempting, therefore, to conclude that when this game is played many times, each player should again randomize his or her decisions and play them each with a frequency of 1/4. This, however, turns out to be a poor strategy. If Ruth did randomize this way, then Charlie could ensure his coming out ahead in the long run by consistently using option (1, 3). Note that in that case the expected outcome per play can be computed by the rules of probability (see addendum at the end of this section) as

$$\frac{1}{4} \cdot 2 + \frac{1}{4} \cdot 0 + \frac{1}{4} \cdot 0 + \frac{1}{4} \cdot (-3) = -\frac{1}{4}$$

In other words, if Ruth plays each option with a frequency of 1/4 and Charlie consistently uses option (1, 3), Charlie can expect to win, on the average, 1/4 of a penny per play. Thus, this game differs drastically from the two previous ones. Whereas the symmetry of the game makes it clear that each player should be able to pursue a long-term strategy that entitles them to expect to come out more or less even in the long run, it is not at all clear this time what this strategy is. We return to this game and issue in Exercise 4.5.50.

The above considerations point out a difficulty that will have to be dealt with when the general theory of games is proposed. It was noted that if Ruth randomizes her behavior by using each of her four options with a frequency of 1/4, then Charlie can guarantee a long-run advantage by consistently using option (1, 3). However, should Charlie chose to do so, Ruth is bound to notice his bias, and she will in all likelihood respond by consistently opting for (1, 2), yielding her a win of 2 on each play. Charlie, then, will respond by playing (2, 3) consistently for a win of 3 on each play. Ruth, then, will switch to (2, 4) for a repeated gain of 4. Charlie, then, will switch back to (1, 3), thus beginning the whole cycle again. A reasonable theory of games should provide stable strategies that avoid such "logical loops," and we will see that such stability can indeed be provided.

BOMBING SORTIES.   *Ruth and Charlie are generals of opposing armies. Every day Ruth sends out a bombing sortie that consists of a heavily armed bomber plane and a lighter support plane. The sortie's mission is to drop a single bomb on Charlie's forces. However, a fighter plane of Charlie's is waiting for them in ambush and it dives down and attacks one of the planes in the sortie once. The bomber has an 80% chance of surviving such an attack, and if it survives it is sure to drop the bomb right on the target. General Ruth also has the option of placing the bomb on the support plane. In that case, due to this plane's lighter armament and lack of proper equipment, the bomb will reach its target with a probability of only 50% or 90%, depending on whether or not it is attacked by Charlie's fighter.*

This information is summarized below where the entries denote the probability of the bomb's delivery:

		Attack	
		Bomber	Support
Bomb	Bomber	80%	100%
placement	Support	90%	50%

Ruth knows that if the bomb is placed consistently on the bomber, she can reasonably expect at least 80% of the missions to succeed. In all likelihood, Charlie's observers at the bombing site would notice this bias and he would direct his fighter plane pilot to always attack the bomber, thus holding Ruth's expectation down to 80% and no more. However, Ruth, who is an experienced poker player, decides to bluff by placing the bomb on the support plane occasionally. Let us say for the sake of argument that she does so $1/4 = 0.25$ of the time. Charlie now faces a dilemma. His observers have advised him of Ruth's new strategy and he suspects that it would be advantageous for him to attack the support plane some of the time, but how often should he do so?

Suppose Charlie decides to counter Ruth's bluffing by attacking the support plane half the time. In this case the situation is summarized as follows:

Attack frequencies

0.50   0.50

Bomber   Support

	Bomber	Support
Bomb placement	80%	100%
frequencies	90%	50%

Because Ruth and Charlie make their daily decisions independently of each other, it follows that on any single sortie, the probabilities of each of the four possible outcomes occurring are as displayed in Table 4.1. Hence, under these circumstances, wherein the bomb is placed on the support plane $1/4$ of the time and this plane is attacked by the fighter $1/2$ of the time, the percentage of successful missions, as computed on the basis of Table 4.1, is

$$0.375 \cdot 80\% + 0.375 \cdot 100\% + 0.125 \cdot 90\% + 0.125 \cdot 50\%$$
$$= 30\% + 37.5\% + 11.25\% + 6.25\% = 85\%$$

Compound event	Probability of event	Likelihood of success of sortie
The bomb is on the bomber *and* the fighter attacks the bomber	$0.75 \cdot 0.50 = 0.375$	80%
The bomb is on the bomber *and* the fighter attacks the support plane	$0.75 \cdot 0.50 = 0.375$	100%
The bomb is on the support plane *and* the fighter attacks the bomber	$0.25 \cdot 0.50 = 0.125$	90%
The bomb is on the support plane *and* the fighter attacks the support plane	$0.25 \cdot 0.50 = 0.125$	50%

Table 4.1   Bombing sorties, first scenario

Thus, this response of Charlie's to Ruth's bluffing has created a situation wherein the bomb can be expected to get through 85% of the time. Because this figure amounts to only 80% when the bomb is placed consistently in the bomber, Ruth's bluffing seems to have paid off.

However, Charlie can change his response pattern. He could, say, decide to diminish the frequency of attacks on the support plane to only 1/5 of the time, leading to the following situation:

			Attack frequencies	
			0.80	0.20
			Bomber	Support
Bomb placement frequencies	0.75	Bomber	80%	100%
	0.25	Support	90%	50%

In this case the percentage of successful sorties, computed on the basis of Table 4.2, is

$$0.60 \cdot 80\% + 0.15 \cdot 100\% + 0.20 \cdot 90\% + 0.05 \cdot 50\%$$
$$= 48\% + 15\% + 18\% + 2.5\% = 83.5\%$$

Thus, by diminishing the frequency of attacks on the support plane, Charlie would create a situation wherein only 83.5% of the sorties would be successful. From Charlie's

Compound event	Probability of event	Likelihood of success of sortie
The bomb is on the bomber *and* the fighter attacks the bomber	$0.75 \cdot 0.80 = 0.60$	80%
The bomb is on the bomber *and* the fighter attacks the support plane	$0.75 \cdot 0.20 = 0.15$	100%
The bomb is on the support plane *and* the fighter attacks the bomber	$0.25 \cdot 0.80 = 0.20$	90%
The bomb is on the support plane *and* the fighter attacks the support plane	$0.25 \cdot 0.20 = 0.05$	50%

Table 4.2   Bombing sorties, second scenario

point of view this is an improvement on the 85% computed previously. Could greater improvements be obtained by a further diminution of the frequency of attacks on the support plane? Suppose these attacks are completely eliminated. Then the state of affairs is as follows:

		Attack frequencies	
		1	0
		Bomber	Support
Bomb placement	0.75 Bomber	80%	100%
frequencies	0.25 Support	90%	50%

and the associated Table 4.3 tells us that the percentage of successful sorties is

$$0.75 \cdot 80\% + 0 \cdot 100\% + 0.25 \cdot 90\% + 0 \cdot 50\%$$
$$= 60\% + 0\% + 22.5\% + 0\% = 82.5\%$$

This would seem to indicate that when Ruth bluffs 1/4 of the time, Charlie should nevertheless ignore the support plane and direct his attacks exclusively at the bomber. Moreover, these calculations indicate that this strategy enables Ruth to improve on her bomber's 80% delivery rate by another 2.5%.

These various analyses raise several questions that guide the development of the subsequent sections. Can Ruth improve on the above 82.5% with a different strategy? What is the best improvement Ruth can obtain? What is Charlie's best response to any specific strategy of Ruth's? Does Charlie have an overall best strategy that is independent of Ruth's decisions?

Compound event	Probability of event	Likelihood of success of sortie
The bomb is on the bomber *and* the fighter attacks the bomber	$0.75 \cdot 1 = 0.75$	80%
The bomb is on the bomber *and* the fighter attacks the support plane	$0.75 \cdot 0 = 0$	100%
The bomb is on the support plane *and* the fighter attacks the bomber	$0.25 \cdot 1 = 0.25$	90%
The bomb is on the support plane *and* the fighter attacks the support plane	$0.25 \cdot 0 = 0$	50%

Table 4.3   Bombing sorties, third scenario

4.1.1   An Addendum on Probabilistic Matters

Unless otherwise stated, it is always assumed here that each time a game is played Ruth and Charlie make their respective decisions independently of each other. This comes under the general heading of *Independent Random Events*, that is, random events whose outcomes have no bearing on each other. For example, if a nickel and a dime are tossed, the outcomes are in general independent, unless the two coins are glued to each other. Similarly, suppose Ruth draws a card at random from a standard deck, replaces it, shuffles the deck, and then Charlie draws a card from that deck. The two draws are then independent. On the other hand, if Ruth does not replace her card in the deck, then Charlie's draw is very much affected by Ruth's draw, because he cannot possibly draw the same card as she did. In this case the two random draws are not independent. The probabilities of independent events are related by the following formula:

If the two events E and F are independent then
probability of $(E \text{ and } F) = (\text{probability of } E) \cdot (\text{probability of } F)$

Thus, the probability of the aforementioned coins both coming up *heads* is $0.5 \cdot 0.5 = 0.25$. Similarly, if a nickel and a standard six-sided die are tossed simultaneously, then the probability of the coin coming up *tails* and the die showing a 3 is $\frac{1}{2} \cdot \frac{1}{6} = \frac{1}{12}$.

Some random events have numerical values attached to their outcomes. Thus, the faces of the standard die have dots marked on them, a person selected at random has a height, and a lottery ticket has a monetary value (that is unknown at the time it is purchased). Loosely speaking, the *expected value* is the average of those numerical values when they are weighted by the corresponding probabilities. More formally,

If the random variable X assumes the numerical values
$x_1, x_2, ..., x_n$ with probabilities $p_1, p_2, ..., p_n$, respectively, then the
expected value (weighted average) of X is
$$p_1 x_1 + p_2 x_2 + \cdots + p_n x_n$$

For example, suppose that the records of an insurance company indicate that during a year they will pay out for accidents according to the following pattern:

$100,000 with probability 0.0002
$50,000 with probability 0.0015
$25,000 with probability 0.003
$5000 with probability 0.01
$1000 with probability 0.03
$0 with probability 0.9553

Then the expected payment per car during the next year is

$$\$100,000 \cdot 0.0002 + \$50,000 \cdot 0.0015 + \$25,000 \cdot 0.003 + \$5000 \cdot 0.01$$
$$+ \$1000 \cdot 0.03 + \$0 \cdot 0.9553 = \$250$$

Similarly, if 5000 lottery tickets are sold of which one will win $10,000, two will win $1000, five will win $100, and the rest will all receive a consolation prize worth one dime, then the expected value of each ticket is

$$\$10,000 \cdot \frac{1}{5000} + \$1000 \cdot \frac{2}{5000} + \$100 \cdot \frac{5}{5000} + \$.10 \cdot \frac{4992}{5000} \approx \$2.60$$

EXERCISES 4.1

1. Ruth is playing Morra and has decided to play the options with the following frequencies:

(1, 2)	-	40%
(1, 3)	-	30%
(2, 3)	-	20%
(2, 4)	-	10%

    a) Design a spinner that facilitates this pattern.

    b) What is the expected outcome if Morra consistently plays (1, 2)?

    c) What is the expected outcome if Morra consistently plays (1, 3)?

    d) What is the expected outcome if Morra consistently plays (2, 3)?

    e) What is the expected outcome if Morra consistently plays (2, 4)?

    f) Which of the above consistent moves is best for Morra?

2. Ruth is playing Morra and has decided to play the options with the following frequencies:

(1, 2)	-	0%
(1, 3)	-	50%
(2, 3)	-	50%
(2, 4)	-	0%

    a) Design a spinner that facilitates this pattern.

    b) What is the expected outcome if Morra consistently plays (1, 2)?

    c) What is the expected outcome if Morra consistently plays (1, 3)?

    d) What is the expected outcome if Morra consistently plays (2, 3)?

    e) What is the expected outcome if Morra consistently plays (2, 4)?

    f) Which of the above consistent moves is best for Morra?

3. Ruth is playing Morra and has decided to play the options with the following frequencies:

$$(1, 2) \quad - \quad 10\%$$
$$(1, 3) \quad - \quad 20\%$$
$$(2, 3) \quad - \quad 30\%$$
$$(2, 4) \quad - \quad 40\%$$

a) Design a spinner that facilitates this pattern.

b) What is the expected outcome if Morra consistently plays $(1, 2)$?

c) What is the expected outcome if Morra consistently plays $(1, 3)$?

d) What is the expected outcome if Morra consistently plays $(2, 3)$?

e) What is the expected outcome if Morra consistently plays $(2, 4)$?

f) Which of the above consistent moves is best for Morra?

4. General Ruth has decided that she will put the bomb on the support plane 10% of the time.

a) Design a spinner that facilitates this pattern.

b) What is the expected mission success rate if General Charlie persists in attacking the bomber exclusively?

c) What is the expected mission success rate if General Charlie persists in attacking the support plane exclusively?

d) What is the expected mission success rate if General Charlie attacks each plane 50% of the time?

e) What is the expected mission success rate if General Charlie attacks the support plane 40% and the bomber 60% of the time?

5. General Ruth has decided that she will put the bomb on the support plane 30% of the time.

a) Design a spinner that facilitates this pattern.

b) What is the expected mission success rate if General Charlie persists in attacking the bomber exclusively?

c) What is the expected mission success rate if General Charlie persists in attacking the support plane exclusively?

d) What is the expected mission success rate if General Charlie attacks each plane 50% of the time?

e) What is the expected mission success rate if General Charlie attacks the support plane 40% and the bomber 60% of the time?

6. General Ruth has decided that she will put the bomb on the support plane 40% of the time.

   a) Design a spinner that facilitates this pattern.

   b) What is the expected mission success rate if General Charlie persists in attacking the bomber exclusively?

   c) What is the expected mission success rate if General Charlie persists in attacking the support plane exclusively?

   d) What is the expected mission success rate if General Charlie attacks each plane 50% of the time?

   e) What is the expected mission success rate if General Charlie attacks the support plane 40% and the bomber 60% of the time?

4.2   Formal Zero-Sum Games

It is time to make some formal definitions. For purely pedagogical reasons we begin with games in which each of the players has only two options, leaving the more general case for the end of the section.

Taking our cue from the penny-matching and bombing sorties games of the previous section, we define a 2×2 *zero-sum game* as a square array of $2 \times 2 = 4$ numbers. Thus, the *mathematical* representation of the penny-matching game is the array

1	−1
−1	1

and the *mathematical* representation of the bombing sorties game is the array

80%	100%
90%	50%

Just as it is convenient to represent the addition of eight oranges to six oranges by the abstract equation $8 + 6 = 14$, we ignore the actual details of the games in most of the subsequent discussion and simply deal with arrays of unitless numbers. Thus, the most general 2×2 zero-sum game has the form

a	b
c	d

where a, b, c, d are arbitrary numbers. This abstraction has the advantages of succinctness and clarity. We, however, make a point of discussing some real games every now and then, and many more such games can be found in the exercises.

Each 2×2 zero-sum game has two players, whom we continue to call Ruth and Charlie. The mathematical analog of deciding on one of the options is the selection of either a row or a column of this array. Specifically, Ruth decides on an option by selecting a row of the array, whereas Charlie makes his decision by specifying a column. Thus, in bombing sorties, Ruth's placement of the bomb in the bomber and Charlie's attacking the support plane are tantamount to Ruth's selecting the first row and Charlie's selecting the second column of the array

80%	100%
90%	50%

Each individual *play* of the game consists of such a pair of selections, made simultaneously and independently. The selected row and column constitute the *outcome* of the play, and its *payoff* is the entry of the array that is contained in both of the selections. Thus, the payoff of the play illustrated above is 100%. On the other hand, had Ruth selected the second row and had Charlie stayed with the second column, the payoff would have been 50%. This payoff of course represents Ruth's winnings (and Charlie's loss) from that play, and she will in general wish to maximize its value, whereas Charlie will be guided by the desire to minimize this payoff.

Informally speaking, a players' strategy is a decision on the frequency with which each available option will be chosen. More formally, a *mixed strategy* is a pair of numbers $[a, b]$,

$$0 \le a \le 1, \quad 0 \le b \le 1, \quad a + b = 1$$

where a denotes the frequency with which the first row (or column) is chosen and b denotes the frequency with which the second row (or column) is chosen. Thus, Ruth's decision to place the bomb on the support plane 1/4 of the time is denoted by the pair $[0.75, 0.25]$, and Charlie's strategy of attacking the two planes with equal frequencies is denoted by $[0.5, 0.5]$. When discussing general strategies, it is convenient to denote Ruth's strategy by $[1 - p, p]$ and Charlie's strategy by $[1 - q, q]$.

This formalization of the intuitive concept of strategy as a list of probabilities was first offered by the mathematician Emile Borel (1871–1956) in a series of papers written in the 1920s. Borel was also the first one to view these games as rectangular arrays. The decision to limit the notion of strategy in this manner could not have been an easy one, and Borel's papers exhibit some ambivalence on this issue. It is very tempting to believe that one can gain some advantage over the opponent by varying one's probabilities at each play, as did Borel, but it is not at all clear how to formulate such a variation. At the conclusion of one of his papers he wrote the following:

> The function $\phi(x, y)$ [the strategy] must then vary at each instant, and vary without following any law at all. One may well doubt if it is possible to indicate an effective and sure means of carrying out such counsel. It seems that, to follow it to the letter, a complete incoherence of mind would be needed, combined, of course, with the intelligence necessary to eliminate those methods we have called bad.

It is our contention, and assumption, that, in practice, every player has a strategy in the sense of a pair of probabilities. Truly random behavior is impossible to attain within the context of a game. For how can any person repeatedly play a game such as rock-scissors-paper in a truly random fashion, that is to say without some pattern of frequencies emerging from his choices? If he makes each individual decision in his head, his past experience and personal preferences are sure to dictate a pattern. If, on the other hand, he uses some device such as a die or a spinner to implement the randomization, the device itself shapes the randomness into a frequency distribution. Each face of the die will come up approximately 1/6 of the time and the spinner's arrow will stop within each sector with a frequency that is proportional to that sector's central angle.

Another argument against the feasibility of a strategyless manner of playing is that each actual sequence of choices made by a player over her lifetime can be construed as a strategy. If a player of penny matching has been observed to have displayed *heads* 83 times and *tails* 39 times, it could be reasonably said that because 83 + 39 = 122, it follows that this player has the strategy

$$\left[\frac{83}{122}, \frac{39}{122}\right] \approx [0.68, 0.32]$$

In reality, of course, such statistics are rarely available. However, the point we are making here is that it is permissible to assume that *every* player is indeed using some strategy that may or may not be known to the opponent.

More serious than the inability of people to behave in a truly random fashion is the theoretical impossibility of treating such a concept within a mathematical framework. Even in the mathematical theory of statistics, where random variables are as commonplace as ants at a picnic, every such variable is assumed to have a probability distribution. Von Neumann and Morgenstern, who also identified strategies with probabilities as was done above, explicitly admitted that they were doing so simply because they had no theory that could handle anything else.

Thus, it will henceforth be assumed that

Every player uses a strategy.

It was seen in bombing sorties that the specification of each player's strategy resulted in a situation wherein an expected payoff could be computed. This *expected payoff* can be defined and computed for arbitrary 2×2 games in a similar manner. Thus, given the strategies $[1 - p, p]$ and $[1 - q, q]$ for the general 2×2 zero-sum game,

	$1 - q$	q
$1 - p$	a	b
p	c	d

(4.1)

the likelihood of Ruth getting the payoff a is the probability of her choosing the first row and Charlie choosing the first column. Because these choices are made independently, we conclude that

the probability of Ruth getting payoff a is $(1 - p)(1 - q)$

Similarly,

the probability of Ruth getting payoff b is $(1 - p)q$

the probability of Ruth getting payoff c is $p(1 - q)$

the probability of Ruth getting payoff d is pq

Because these four events are mutually exclusive and they exhaust all the possibilities, it follows that Ruth's expected payoff is

$$(1 - p)(1 - q)a + (1 - p)qb + p(1 - q)c + pqd$$

Diagrams such as that of (4.1) above, wherein the players' strategies are appended to the game's array, are called *auxiliary diagrams*. They turn the computation of the expected payoff into a routine task and so are repeatedly used in the sequel as a visual aid.

EXAMPLE 4.2.1. *For the bombing sorties game, Ruth's strategy [0.3, 0.7] and Charlie's strategy [0.6, 0.4] yield the auxiliary diagram*

	0.6	0.4
0.3	80%	100%
0.7	90%	50%

The corresponding payoff is

$$0.3 \cdot 0.6 \cdot 80\% + 0.3 \cdot 0.4 \cdot 100\% + 0.7 \cdot 0.6 \cdot 90\% + 0.7 \cdot 0.4 \cdot 50\%$$
$$= 14.4\% + 12\% + 37.8\% + 14\% = 78.2\%$$

In other words, when the players use the above specified strategies, Ruth can expect 78.2% of the missions to be successful.

EXAMPLE 4.2.2. *Compute the expected payoff of the strategies [0.2, 0.8] of Ruth and [0.3, 0.7] of Charlie for the abstract 2×2 zero-sum game:*

5	0
-1	2

The auxiliary diagram is

	0.3	0.7
0.2	5	0
0.8	−1	2

and so the expected payoff is

$$0.2 \cdot 0.3 \cdot 5 + 0.2 \cdot 0.7 \cdot 0 + 0.8 \cdot 0.3 \cdot (-1) + 0.8 \cdot 0.7 \cdot 2 =$$
$$0.3 + 0 - 0.24 + 1.12 = 1.18$$

In other words, when the players use the specified strategies within the context of the given (repeated) game, Ruth can expect to win, on the average, 1.18 per play. Of course, this being a zero-sum game, Charlie should expect to lose the same amount per play.

We now turn to the formal definition of games in which one or both of the players have more than two options. If m and n are any two positive integers, then an $m \times n$ *zero-sum game* is a rectangular array of mn numbers having m rows and n columns. Thus,

0	1	−1
−1	0	1
1	−1	0

is a 3×3 zero-sum game (rock-paper-scissors), and

0	2	−3	0
−2	0	0	3
3	0	0	−4
0	−3	4	0

is a 4×4 zero-sum game (two-finger Morra). The array below is an example of an abstract 3×4 zero-sum game (an interesting concrete nonsquare game is discussed in Section 5 in detail).

2	−1	−5	3
0	−2	3	−3
1	0	1	−2

(4.2)

It is again assumed that in each play Ruth selects a row and Charlie selects a column of the array. The selected row and column constitute the *outcome* of that play, and the entry of the array which is the intersection of Ruth's chosen row with Charlie's chosen column is the corresponding *payoff*. Thus, in the previous abstract game, if Ruth selects the second row and Charlie selects the fourth column, the corresponding payoff is −3, a loss for Ruth.

Given an $m \times n$ zero-sum game, a *strategy* for Ruth is a list of numbers $[p_1, p_2, ..., p_m]$, such that

$$0 \le p_i \le 1 \text{ for each } i = 1, 2, ..., m \quad \text{and} \quad p_1 + p_2 + \cdots + p_m = 1$$

where p_i denotes the frequency with which Ruth chooses the ith row. Similarly, a strategy for Charlie is an ordered list of numbers $[q_1, q_2, ..., q_n]$, such that

$$0 \le q_j \le 1 \text{ for each } j = 1, 2, ..., n \quad \text{and} \quad q_1 + q_2 + \cdots + q_n = 1$$

where q_j denotes the frequency with which Charlie chooses the jth column. Thus, in the symmetric rock-scissors-paper game, the strategy [0.6, 0.3, 0.1] denotes the decision, by either player, to display *rock* 60% of the time, *scissors* 30% of the time, and *paper* only 10% of the time. In the above abstract 3×4 zero-sum game (or in any 3×4 zero-sum game for that matter) the strategy [0.2, 0.3, 0.5] denotes Ruth's choosing the first, second, or third rows 20%, 30%, 50% of the time, respectively. The strategy [0.4, 0, 0.1, 0.5] denotes Charlie's choosing the first, second, third, or fourth columns 40%, 0%, 10%, and 50% of the time, respectively. A *pure strategy* is one that calls for the exclusive use of a particular row or column. In bombing sorties each player has two pure strategies: [1, 0], which calls for the use of the first row or column only, and [0, 1], which calls for the exclusive use of the second row or column. In the abstract game of (4.2), Ruth has three pure strategies: [1, 0, 0], [0, 1, 0], and [0, 0, 1]. In the same game Charlie has four pure strategies: [1, 0, 0, 0], [0, 1, 0, 0], [0, 0, 1, 0], and [0, 0, 0, 1]. Strategies that are not known to be pure are called *mixed*. Thus, the strategies [0.3, 0.7] and $[1 − p, p]$ are mixed strategies, even though the latter may turn out to be pure when p is either 0 or 1.

Every such choice of specific strategies on the part of both players narrows the situation down to a point where an expected payoff can be computed. Suppose Ruth uses the strategy $[p_1, p_2, ..., p_m]$ and Charlie uses the strategy $[q_1, q_2, ..., q_n]$ in some abstract $m \times n$ zero-sum game. If $a_{i,j}$ denotes the payoff in the ith row and the jth column of this game,

then the likelihood of this payoff actually taking place is the probability of Ruth choosing the ith row and Charlie choosing the jth column, which, of course, equals

$p_i q_j$. Thus, this specific outcome's contribution to the expected payoff is $p_i q_j a_{i,j}$. Consequently the total *expected payoff*, being the sum of all these contributions, equals

$$\text{Sum of all } p_i q_j a_{i,j} \quad i = 1, 2, ..., m, \quad j = 1, 2, ..., n$$

EXAMPLE 4.2.3. *Compute the expected payoff when Ruth uses the strategy [0.2, 0.3, 0.5] and Charlie uses the strategy [0.1, 0.7, 0.2] in the rock-scissors-paper game.*

The auxiliary diagram is

	0.1	0.7	0.2
0.2	0	1	−1
0.3	−1	0	1
0.5	1	−1	0

and the corresponding sum is

$$0.2 \cdot 0.1 \cdot 0 + 0.2 \cdot 0.7 \cdot 1 + 0.2 \cdot 0.2 \cdot (-1) +$$
$$0.3 \cdot 0.1 \cdot (-1) + 0.3 \cdot 0.7 \cdot 0 + 0.3 \cdot 0.2 \cdot 1 +$$
$$0.5 \cdot 0.1 \cdot 1 + 0.5 \cdot 0.7 \cdot (-1) + 0.5 \cdot 0.2 \cdot 0 =$$
$$0 + 0.14 - 0.04 - 0.03 + 0 + 0.06 + 0.05 - 0.35 + 0 = -0.17$$

In other words, under these circumstances, Ruth should expect to lose 0.17 pennies per play.

EXAMPLE 4.2.4. *Compute the expected payoff when the strategies [0.3, 0, 0.7] and [0.1, 0.2, 0.3, 0.4] are used by Ruth and Charlie, respectively, in the abstract game (4.2) above.*

Using the auxiliary diagram

	0.1	0.2	0.3	0.4
0.3	2	−1	−5	3
0	0	−2	3	−3
0.7	1	0	1	−2

we get an expected payoff of

$$0.3 \cdot 0.1 \cdot 2 + 0.3 \cdot 0.2 \cdot (-1) + 0.3 \cdot 0.3 \cdot (-5) + 0.3 \cdot 0.4 \cdot 3 +$$
$$0 \cdot 0.1 \cdot 0 + 0 \cdot 0.2 \cdot (-2) + 0 \cdot 0.3 \cdot 3 + 0 \cdot 0.4 \cdot (-3) +$$
$$0.7 \cdot 0.1 \cdot 1 + 0.7 \cdot 0.2 \cdot 0 + 0.7 \cdot 0.3 \cdot 1 + 0.7 \cdot 0.4 \cdot (-2) =$$
$$0.06 - 0.06 - 0.45 + 0.36 + 0 + 0 + 0 + 0 + 0.07 + 0 + 0.21 - 0.56 =$$
$$-0.37$$

EXERCISES 4.2

In each of Exercises 1–11 compute the expected payoff where R denotes a strategy for Ruth and C denotes a strategy for Charlie for the given game G.

1. $R = [0.2, 0.8]$, $C = [0.7, 0.3]$, $G =$

2	3
4	1

2. $R = [0.6, 0.4]$, $C = [0, 1]$, $G =$

2	3
4	1

3. $R = [0.2, 0.8]$, $C = [0.7, 0.3]$, $G =$

−1	3
4	−2

4. $R = [0.6, 0.4]$, $C = [0, 1]$, $G =$

−1	3
4	−2

5. $R = [0.6, 0.3, 0.1]$, $C = [0.1, 0.4, 0.3, 0.2]$, $G =$

1	0	−2	3
−3	4	2	−4
0	−1	0	1

6. $R = [0.6, 0, 0.04]$, $C = [0, 0.5, 0.3, 0.2]$, $G =$

1	0	−2	3
−3	4	2	−4
0	−1	0	1

7. $R = [0, 1, 0]$, $C = [0.5, 0, 0, 0.5]$, $G =$

1	0	−2	3
−3	4	2	−4
0	−1	0	1

8. $R = [0.2, 0, 0.4, 0, 0.4]$, $C = [0.1, 0.1, 0.8]$, $G =$

1	−3	2
0	4	−4
−2	0	2
3	−3	−1
−3	5	1

9. $R = [0, 0, 0, 0.4, 0.6]$, $C = [0.1, 0, 0.9]$, $G =$

1	-3	2
0	4	-4
-2	0	2
3	-3	-1
-3	5	1

10. $R = [0, 0, 0, 1, 0]$, $C = [0, 1, 0]$, $G =$

1	-3	2
0	4	-4
-2	0	2
3	-3	-1
-3	5	1

11. $R = [0, 1, 0, 0, 0]$, $C = [0, 0, 1]$, $G =$

1	-3	2
0	4	-4
-2	0	2
3	-3	-1
-3	5	1

Describe the following situations as mathematical zero-sum games.

12. *The River Tale* (J. D. Williams, 1986). Steve is approached by a stranger who suggests that they match coins. Steve says that it's too hot for violent exercise. The stranger says, "Well then, let's just lie here and speak the words *heads* or *tails*—and to make it interesting I'll give you $30 when I call *tails* and you call *heads* and $10 when it's the other way round. And—just to make it fair—you give me $20 when we match."

13. *The Birthday* (J. D. Williams, 1986). Frank is hurrying home late, after a particularly grueling day, when it pops into his mind that today is Kitty's birthday! Or is it? Everything is closed except the florist. If it is not her birthday and he brings no gift, the situation will be neutral, i.e., payoff 0. If it is not and he comes in bursting with roses and obviously confused, he may be subjected to the Martini test, but he will emerge in a position of strong one-upness, which is worth 1. If it is her birthday and he has, clearly, remembered it, that is worth somewhat more, say 1.5. If he has forgotten it he is down like a stone, say, −10.

14. *The Hi-Fi* (J. D. Williams, 1986). The firm of Gunning & Kappler manufactures an amplifier. Its performance depends critically on the characteristics of one small inaccessible condenser. This normally costs Gunning and Kappler $1, but they are set back a total of $10, on the average, if the original condenser is defective. There are some alternatives open to them. It is possible for them to buy a superior quality condenser, at $6, which is fully guaranteed; the manufacturer will make good the condenser and the costs incurred in getting the amplifier to operate. Also available is a condenser covered by an insurance policy that states, in effect, "If it is our fault, we will bear the costs and you get your money back." This item costs $10. (This is a 3×2 game that Gunning & Kappler is playing against Nature, whose options are to supply either a defective or a nondefective condenser.)

15. *The Huckster* (J. D. Williams, 1986). Merrill has a concession at the Yankee Stadium for the sale of sunglasses and umbrellas. The business places quite a strain on him, the weather being what it is. He has observed that he can sell about 500 umbrellas when it rains and about 100 when it shines; in the latter case he also can dispose of about 1000 sunglasses. Umbrellas cost him 50 cents and sell for $1 (this is 1954); glasses cost 20 cents and sell for 50 cents. He is willing to invest $250 in the project. Everything that isn't sold is a total loss (the children play with them).

16. *The Coal Problem* (J. D. Williams, 1986). On a sultry summer afternoon, Hans' wandering mind alights upon the winter coal problem. It takes about 15 tons to heat his house during a normal winter, but he has observed extremes when as little as 10 tons and as much as 20 were used. He also recalls that the price per ton seems to fluctuate with the weather, being $10, $15, and $20 a ton during mild, normal, and severe winters, respectively. He can buy now, however, at $10 a ton. He considers three pure strategies, namely, to buy 10, 15, or 20 tons now and the rest, if any, later. He will be moving to California in the spring and he cannot take excess coal with him.

4.3   Pure Strategies for Nonzero-Sum Games

In the zero-sum games discussed in the previous section the payoffs are numerical, each player's gain is the other player's loss, and the players make their decisions simultaneously and independently of each other. There is, however, a considerable amount of interest in other types of games that may possess only some or even none of these properties: Payoffs may be unquantifiable as is the case when loss of face is at stake, and in most wars both players sustain tremendous losses. Although there are many different generalizations of zero-sum games, the remainder of this chapter is confined to a discussion of *two-person noncooperative nonzero-sum games*. The term *noncooperative* refers to the assumption that the players do not consult each other about ways and means for improving their payoffs. In the interest of brevity, these games are referred to as simply *nonzero-sum games* or even as just *games* henceforth.

Nonzero-sum games are also modeled by having Ruth select a row and Charlie select a column from a suitable rectangular array. However, because a player's gain is not necessarily her opponent's loss anymore, the payoff of each outcome can no longer be described by a single number. The result of Ruth choosing row i and Charlie choosing column j is the *payoff pair*

$$(a_i, b_j)$$

where a_i denotes Ruth's payoff and b_j denotes Charlie's payoff. Thus, the general 2×2 nonzero-sum game has the array

Charlie

Ruth

(a_1, b_1)	(a_2, b_2)
(a_3, b_3)	(a_4, b_4)

(4.3)

It is assumed that despite the presence of two payoffs in each pair, players only take their own payoffs into consideration when deciding on a move. This is commonly formalized as follows.

The Principle of Rationality: *Every player wishes to come out as well off as possible.*

In other words, cutting off one's nose to spite one's spouse is not rational behavior, and such impulses are given no consideration in this text.

One of the most popular factors in the decision-making process is the issue of guarantees. Given a variety of options, what is the best outcome that can be guaranteed? In the context of games, the strategy that reflects this concern is called the *pure maximin* strategy. For Ruth this strategy dictates the choice of the row with the largest minimum first payoff; for Charlie this is the column with the largest minimum second payoff. The payoff guaranteed to a player by his maximin strategy is that player's *maximin value*. The payoff pair in the outcome determined by the two pure maximin solutions is called the *pure value pair* of the game. The pure maximin strategies and values as well as the pure value pairs of any nonzero-sum game are easily derived.

EXAMPLE 4.3.1. *Find the pure maximin strategies and values and also the pure value pairs of the game*

$(2, 3)$	$(3, 5)$
$(2, 2)$	$(1, 3)$

To the right of each row, record that row's smallest first entry, and at the bottom of each column, record that column's smallest entry. Circle the largest number on the right and

(2,3)	(3,5)	②
(2,2)	(1,3)	3

2   ③

the largest number at the bottom. The circled items indicate the maximin strategies for both players: row 1 for Ruth and column 2 for Charlie. The value pair is $(3, 5)$ and happens to yield each player a payoff that is better than the guarantee sought.

EXAMPLE 4.3.2. *Find the pure maximin strategies and values and also the pure value pairs of the game*

(5,2)	(3,0)	(8,1)	(2,3)
(6,3)	(5,4)	(7,4)	(1,2)
(7,5)	(4,6)	(6,8)	(0,2)

To the right of each row, record that row's smallest first entry, and at the bottom of each column, record that column's largest entry. Circle the largest number on the right and the

(5,2)	(3,0)	(8,1)	(2,3)	②
(6,3)	(5,4)	(7,4)	(1,2)	1
(7,5)	(4,6)	(6,8)	(0,2)	0

②   0   1   ②

largest numbers on the bottom. Ruth's maximin strategy is therefore row 1. Charlie has two maximin strategies, columns 1 and 4, yielding payoff pairs $(5, 2)$ and $(2, 3)$, respectively. Charlie may prefer to play column 4 because it seems to promise a better payoff of 3, but this is no longer guaranteed. From the point of view of guarantees, the two maximin strategies of column 1 and column 4 have the same value for Charlie.

We now go on to consider several games that are played very frequently and in many guises.

PRISONER'S DILEMMA.   *Two people are arrested and held in connection with a certain robbery. The prosecution only has enough evidence to convict them of the robbery itself, but it is believed that the robbers were actually carrying guns at the time, making them liable to the more severe charge of armed robbery. The prisoners are held in separate cells and cannot communicate with each other. Each of them is offered the same deal: If you testify that your partner was armed but he does not testify against you, your sentence will be suspended while he will spend 15 years in jail. If both of you testify against each other, you will both get 10 years, and if neither of you does so, you both get 5 years in jail. Both prisoners are fully aware that both have been offered the same deal. They are given some time to think the deal over, but neither is aware of his expartner's decision.*

This game is clearly nonzero-sum and can be described as follows:

	Refuse the deal	Accept the deal
Refuse the deal	$(-5, -5)$	$(-15, 0)$
Accept the deal	$(0, -15)$	$(-10, -10)$

where the payoff $-x$ denotes a jail term of x years. Each player's pure maximin strategy is easily derived:

$(-5, -5)$	$(-15, 0)$	-15
$(0, -15)$	$(-10, -10)$	$\widehat{-10}$
-15	$\widehat{-10}$	

This maximin strategy recommends that both prisoners accept the deal offered by the prosecution, thereby guaranteeing that each receives a 10-year jail term. Both experience and experimentation testify to the fact that this is frequently the decision made by people in such circumstances. This is unsettling because if both players had refused the deal, they would have each received a 5-year term, a distinctly better sentence for both.

Let us digress to formalize this vague notion of "unsettling." In any nonzero-sum game, a payoff pair (a, b) is said to be *better than* the payoff (a', b') if either of the following conditions hold:

$$a > a' \text{ and } b \geq b' \quad \text{or} \quad a \geq a' \text{ and } b > b'$$

Thus, $(3, 2)$ is better than each of the payoffs $(1, 2)$, $(3, 1)$, and $(2, 2)$. An outcome of a game is *Pareto optimal* if the game possesses no outcome with a better payoff. In game (4.4) below, the outcomes with payoffs $(2, 3)$ and $(1, 5)$ are both Pareto optimal:

$(2, 3)$	$(1, 5)$
$(2, 2)$	$(0, 3)$

(4.4)

On the other hand, in the game of example (4.3) above, the outcome with payoff (3, 5) is the only Pareto optimal outcome.

It is clear that when the players of a game are treated as a group, Pareto optimal outcomes are desirable. The unfortunate and tragic aspect of the prisoners' dilemma game is that both the strategy recommended by game theory and the observed behavior of people result in the outcome with payoff (−10, −10), which is dominated by (−5, −5) and hence is *not* Pareto optimal.

CHICKEN. *Two adversaries are set on a collision course. If both persist, then a very unpleasant outcome, sometimes mutual annihilation, is guaranteed. If only one of the players swerves away ("chickens out"), he loses the game. If both swerve, the result is a draw.*

Usually, the reward for winning this game is merely a sense of dominance, and swerving only entails loss of face, both equally unquantifiable. The possibility of mutual annihilation when both players persist is what makes this a nonzero-sum situation. In that case neither player gains anything; in fact, quite frequently both stand to lose their lives. We offer no solution and wish to discuss only some paradoxical aspects of the game. For this purpose, the game is made concrete by transferring our attention to the Cuban missile crisis of October 1962, certainly one of best known and most frightening chicken games ever played.

In that month the United States discovered that the USSR was building a missile base in Cuba. As a small Soviet fleet was on its way to the island to bring in supplies, the Americans set up a naval blockade. It is quite likely that if these two fleets had met, the conflict would have escalated into a full-fledged, and probably nuclear, war. In the event, the Soviets blinked, their ships withdrew, and the United States "won." It is our purpose here to discuss a curious wrinkle on this confrontation.

Although it is impossible to realistically quantify the various outcomes of this game, it is possible to introduce some numbers into it by a preferential ranking of the various outcomes. There were four possible outcomes for the Cuban missile crisis:

(US swerves, USSR swerves),   (US persists, USSR swerves)
(US swerves, USSR persists),   (US persists, USSR persists).

The players would rank these outcomes preferentially as follows (1 denoting the least desirable and 4 the most desirable outcomes):

US		USSR
(US persists, USSR swerves)	4	(US swerves, USSR persists)
(US swerves, USSR swerves)	3	(US swerves, USSR swerves)
(US swerves, USSR persists)	2	(US persists, USSR swerves)
(US persists, USSR persists)	1	(US persists, USSR persists).

These preferences are summarized as the following nonzero-sum game:

USSR

		Swerve	Persist
		Swerve	Persist
US	Swerve	(3, 3)	(2, 4)
	Persist	(4, 2)	(1, 1)

Here the "payoff" (i, j) associated with any outcome denotes the fact that the US assigns this outcome the ranking i in its preference list, whereas the USSR assigns it the ranking j. The maximin solution of this game recommends that both players swerve and has payoff $(3, 3)$.

Suppose now that as the crisis was unfolding, President Kennedy was notified that his staff was infiltrated by Soviet spies and that whatever decision he made would immediately be made known to Premier Khrushchov. One would believe this to be a cause for much consternation for the Americans. We argue that the opposite is the case and that in fact

> In the game of chicken, a player's foreknowledge of her opponent's decision works to that player's disadvantage, provided that the opponent is aware of the player's foreknowledge.

This conclusion is drawn on the basis of the Principle of Rationality. Taking the infiltration into account, President Kennedy would have reasoned as follows:

> If I decide to <u>swerve</u>, Khrushchov will know it and, choosing between $(3, 3)$ and $(2, 4)$, he will decide to persist, an outcome with payoff $(2, 4)$.

> If I decide to <u>persist</u>, Khrushchov will know it and, choosing between $(4, 2)$ and $(1, 1)$, he will decide to swerve, an outcome with payoff $(4, 2)$.

Thus, if Kennedy swerves, the outcome is $(2, 4)$, whereas if he persists the outcome is $(4, 2)$. He would therefore choose to persist, resulting in the payoff pair $(4, 2)$ wherein the Soviets swerve.

This is, of course, exactly what transpired. On October 25, a Russian convoy headed on a collision course with the American blockade dispersed. It could be argued that although the USSR had in all likelihood not been privy to the American decision-making process, its leaders were wiser and more cognizant of the possible disastrous consequences of political games of chicken. After all, only 17 years had passed since the conclusion of World War II in which 20 million Russians are estimated to have perished and major portions of their land were devastated, whereas not a single one of that war's battles was fought on American soil.

The essential feature of the infiltration that was hypothesized in the Cuban Missile Crisis is that the USSR did not select its option until *after* the US had done so. In general, this prerogative may or may not be beneficial to the player who possesses it.

EXAMPLE 4.3.3. *Consider the game*

(5,2)	(3,0)	(8,1)	(2,3)
(6,3)	(5,4)	(7,4)	(1,1)
(7,5)	(4,6)	(6,8)	(0,2)

Suppose first that Charlie is informed of Ruth's decision when he makes his own, and that Ruth is aware of this fact. Then Ruth will reason as follows:

If I select row 1, Charlie will opt for the payoff pair (2, 3).

If I select row 2, Charlie will opt for one of the payoff pairs (5, 4) and (7, 4).

If I select row 3, Charlie will opt for the payoff pair (6, 8).

The maximin strategy then will direct Ruth to select row 3 and Charlie to choose column 3, resulting in an outcome with payoff pair (6, 8).

On the other hand, if Ruth is the one who is informed of Charlie's decision when she makes her own and if Charlie is aware of this fact, then he will reason as follows:

If I select column 1, Ruth will opt for payoff pair (7, 5).

If I select column 2, Ruth will opt for payoff pair (5, 4).

If I select column 3, Ruth will opt for payoff pair (8, 1).

If I select column 4, Ruth will opt for payoff pair (2, 3).

The maximin strategy then will direct Charlie to choose column 1 and Ruth to choose row 3 for a payoff pair of (7, 5).

If neither player is informed of their opponent's decisions, then Ruth's maximin strategy calls for the selection of row 1 for a guaranteed payoff of at least 2. Charlie's maximin strategy calls for the selection of column 1 for a guaranteed payoff of at least 2. The actual outcome will then have payoff pair (5, 2). Thus, in this game, information about the opponent's decisions is advantageous to each player.

It was seen in the discussion of the chicken game that the maximin analysis can fail to properly describe human behavior in some gamelike situations. In retrospect, in that particular case, the failure was due to the oversimplification that occurred when it was assumed that the players made their decision independently and simultaneously. In most instances of this game, however, the players' behaviors very much affect each other. They continuously observe each other, and when one swerves the other will most certainly persist to the end. Rather than completely disqualify the maximin strategy for such games, a way was found to adapt it so as to continue to supply researchers with satisfactory solutions.

One of the features of the maximin strategy is that although it may not lead to the best of all possible payoffs, it does offer the players the consolation of having eliminated the possibility of their regretting their decisions. Each player knows that he or she has played so as to obtain the best possible guarantees, regardless of what the opponent did. This *No Regrets* policy can be applied to nonzero-sum games as well.

In a nonzero-sum game, an outcome is said to be a pure *Nash equilibrium point* if its payoff pair (a_i, b_j) is such that a_i is maximum for its column and b_j is maximum for its row. Such a Nash equilibrium point has the desired *No Regrets* property because given that Charlie selected column j, Ruth has no reason to regret having selected row i—no other selection of Ruth's would have yielded her a better payoff. Similarly, given that Ruth selected row i, Charlie has no reason to regret having selected column j—no other selection of his would have yielded him a better payoff. Note that in this definition of the Nash equilibrium, each player is assumed to be concerned with regrets over her or his own actions only. Any feelings that they might entertain regarding the other players are ignored. The reason for this limitation is that the incorporation of such feelings into the considerations would result in too many complications.

EXAMPLE 4.3.4. *Find the Nash equilibrium points of the game*

(5,2)	(3,0)	(8,1)	(2,3)
(6,3)	(5,4)	(7,4)	(1,2)
(7,5)	(4,6)	(6,8)	(0,2)

One method for finding these points is to examine all the payoff pairs (a_i, b_j) in succession and to star both a_i that is a maximum for its column and each b_j that is a maximum for its row. The Nash equilibrium points are those whose payoff pairs have *both* entries starred:

(5,2)	(3,0)	(8*,1)	(2*,3*)
(6,3)	(5*,4*)	(7,4*)	(1,2)
(7*,5)	(4,6)	(6,8*)	(0,2)

Thus, the Nash equilibrium points of the given game are those with payoffs (5, 4) and (2, 3). Note that one of these payoff pairs, namely (5, 4), is clearly preferable to the other. Nevertheless, this does not necessarily mean that either Ruth or Charlie will aim for this "better" payoff pair. After all, by selecting the second row [that of the pair (5, 4)], Ruth opens herself up to the possibility of obtaining a payoff of only 1 if Charlie selects column 4. Similarly, if Charlie were to select the second column [that of the pair (5, 4)], he would be opening himself up to the possibility of gain 0 if Ruth selects row 1.

The next example shows that pure Nash equilibrium points need not exist.

EXAMPLE 4.3.5. *Find the Nash equilibrium points of the game*

(2,1)	(1,2)
(1,2)	(2,1)

When the row maximum of the first entry and the column maximum of the second entry of each payoff pair are starred, we obtain the pattern

(2*,1)	(1,2*)
(1,2*)	(2*,1)

Because no payoff pair has both of its entries starred, it follows that this game has no pure Nash equilibrium points.

The logic that underlies the Nash equilibrium is the expectation that as the players watch each other maneuver, the situation will naturally gravitate toward a Nash equilibrium outcome. Such was the case in the Cuban Missile Crisis. As those October days passed, the Russians became convinced of the Americans' determination to persist and so they swerved. The diagram below demonstrates that this actual outcome is indeed one of the two Nash equilibrium points of that confrontation.

USSR

		Swerve	Persist
US	Swerve	(3, 3)	(2*, 4*)
	Persist	(4*, 2*)	(1, 1)

Prisoner's dilemma has only one Nash equilibrium point with payoff pair (−10, −10),

	Refuse the deal	Accept the deal
Refuse the deal	(−5, −5)	(−15, 0*)
Accept the deal	(0*, −15)	(−10*, −10*)

which is consistent both with its maximin strategy and many observed solutions. Here, of course, the players cannot watch each other reach their decisions. Nevertheless, the oft reached outcome wherein both prisoners accept the deal does turn out to be a pure Nash equilibrium point.

The Nash equilibrium points have become a popular tool for theoretical econo-mists. The next example should give the reader some feel for how this concept is used by them.

THE JOB APPLICANTS.   *Firms 1 and 2 have one opening each for which they offer salaries 2a and 2b, respectively (the 2 is only used to prevent fractions from appear-ing later). Ruth and Charlie can each apply to only one of the positions, and they must simultaneously decide whether to apply to firm 1 or to firm 2. If only one of them applies for a job, he or she gets it; if both apply for the same position, the firm hires one of them at random.*

This situation is modeled as the 2×2 nonzero-sum game

Charlie applies to

		Firm 1	Firm 2
Ruth applies	Firm 1	(a, a)	$(2a, 2b)$
to	Firm 2	$(2b, 2a)$	(b, b)

In this array the entries $(2a, 2b)$ and $(2b, 2a)$ are self-explanatory. The entry (a, a) is obtained by reasoning as follows. If both Ruth and Charlie apply to the same firm, then, because it was stipulated that the firm will select an applicant at random, each can expect a payoff of half the salary, i.e., a. A similar line of thought justifies the entry (b, b).

Not surprisingly, the analysis of the game depends on the relationship between a and b. Suppose first that

$$a \le 2b \quad \text{and} \quad b \le 2a$$

In other words, suppose first that the two salaries are not too far out of line with each other in that neither exceeds double the other. Then the two entries $(2a, 2b)$ and $(2b, 2a)$ are both pure Nash equilibria. Such, for example, might be the outcome if Ruth and Charlie became aware of each other's existence and came to some mutual agreement.

On the other hand, if the salaries are out of line with each other, say

$$b > 2a$$

then it is the entry (b, b) that constitutes the unique pure Nash equilibrium point. This corresponds to both Ruth and Charlie applying for the better job. The case where $b = 2a$ is intermediary and has all the above three outcomes as its Nash equilibria.

Thus, this highly simplified model predicts that if the disparity between the salaries is not too great, then some reasonable distribution of the positions may hap-pen. If one job is much better than the other, then both will apply for it, resulting in a situation where one of them remains unemployed.

This model is discussed again in greater depth in the next two sections.

We mention in closing this section that the 2×2 nonzero-sum games have been classified in Rapoport, Guyer, and Gordon (1976) into 78 different types. This classification depends on the distribution of Pareto optimal outcomes, Nash equilibrium points, and their relationship to the maximin strategies.

EXERCISES 4.3

For each of the nonzero-sum games in Exercises 1–10 find the following:

a) All the pure maximin strategies.

b) The pure maximin values.

c) All the pure value pairs.

d) All the Pareto optimal payoffs.

e) All the pure Nash equilibrium points.

f) The outcome of the game if Charlie is aware of Ruth's decision when he makes his, and Ruth knows of this.

g) The outcome of the game if Ruth is aware of Charlie's decision when she makes hers, and Charlie knows of this.

1.

(2,3)	(1,4)
(0,5)	(4,1)

2.

(0,5)	(2,3)
(4,1)	(1,4)

3.

(1,3)	(3,1)
(2,2)	(2,5)

4.

(3,1)	(2,5)
(1,1)	(2,2)

5.

(5,4)	(4,1)
(1,4)	(3,2)
(1,1)	(5,5)

6.

(1,2)	(3,4)
(5,6)	(7,8)
(7,6)	(5,5)

7.

(9,0)	(8,1)	(7,2)
(6,3)	(5,4)	(4,5)
(3,6)	(2,7)	(1,9)

8.

(9,0)	(8,−1)	(7,−2)
(6,−3)	(5,−4)	(4,−5)
(3,−6)	(2,−7)	(1,−9)

9.

(1,2)	(1,−2)	(1,1)	(0,3)
(2,1)	(0,0)	(−3,0)	(1,−2)
(3,−1)	(−2,1)	(2,1)	(2,2)
(2,1)	(2,2)	(1,−1)	(−1,1)

10.

(−1,2)	(−2,1)	(2,0)	(1,2)
(1,1)	(3,0)	(0,−3)	(−2,1)
(1,−2)	(3,0)	(1,2)	(0,3)
(3,0)	(2,2)	(−1,1)	(−1,1)

Describe the following situation as nonzero sum games.

11. *Leader* (A. Coleman). Two motorists are waiting to enter a heavy stream of traffic from opposite ends of an intersection, and both are in a hurry to get to their destinations. When a gap in the traffic occurs, each must decide whether to concede the right of way to the other or to drive into the gap. (Use the preferential ranking of the game of chicken to assign numerical values to the payoffs.)

12. *Battle of the sexes* (Luce and Raiffa). Ruth and Charlie, who are happily married, are planning an evening's entertainment. Ruth would like to go to the concert at the Arts Center, and Charlie would rather stay home and watch the ball game on TV. Still, both would prefer to spend their time together rather than separately. (Use the preferential ranking of the game of chicken to assign numerical values to the payoffs.)

13. *Two leagues* (Dixit & Nalebuff). Two football leagues, USFL and NFL, are deciding whether to schedule their games in the fall or in the spring. They estimate that 10 million viewers will watch football in the fall, but only 5 million will watch in the spring. If one league has a monopoly during a season, it gets the entire market. If both leagues schedule their games for the same season, the NFL gets 70% and the USFL gets 30% of the market.

14. *Oil cartel* (Dixit & Nalebuff). Ruth and Charlie are the rulers of two countries that have formed an oil cartel. To keep the price of oil up, they have agreed to limit their productions, respectively, to 4 million and 1 million barrels per day. For each, cheating means producing an extra 1 million barrels each day. Depending on their decisions, their total output would therefore be 5, or 6, or 7 million barrels, with corresponding profit margins of $16, $12, and $8 per barrel, respectively.

15. *Home defense* (Dixit & Nalebuff). Ruth is a typical homeowner and Charlie an average burglar. Ruth is trying to decide whether to keep a gun in her home, and Charlie faces the options of whether or not to bring a gun to his next break-in. (Use preferential ranking.)

16. *Altruist's dilemma* (Heckathorn). Ruth and Charlie, who are married, are considering their Christmas gift strategy. To simplify matters, suppose they can each either spend a lot or a reasonable amount on each other's presents. Use preferential ranking to display this as a nonzero-sum game.

17. *Assurance game* (Heckathorn). Ruth and Charlie are the employees of a firm that has been remarkably successful over the last 2 years. They know that their boss has more than enough money to give them both a raise. Suppose each has the options of either not doing anything at all or else presenting their boss with the ultimatum: "If you don't give me a raise, I quit!" Use preferential ranking to present this situation as a nonzero-sum game.

18. *Privileged game* (Heckathorn). This is essentially the same as Exercise 17 with the additional wrinkle that Ruth and Charlie are aware that when the boss caves in to one employee's ultimatum, he will automatically give the other employee a smaller raise.

19. Does every nonzero-sum game have to have at least one Pareto optimal outcome? Justify your answer.

20. Give an example of a nonzero-sum game in which all the payoffs are distinct and each outcome is Pareto optimal.

4.4 Mixed Strategies for Nonzero-Sum Games

Just like their zero-sum kin, nonzero-sum games may be repeated and their strategies may be mixed. As was done in the discussion of zero-sum games, we begin with 2×2 nonzero-sum games. A *mixed strategy* for the general 2×2 game is a designation of the frequencies with which each option is exercised and consists of a pair $[a, b]$

Charlie

	Charlie	
Ruth	(a_1, b_1)	(a_2, b_2)
	(a_3, b_3)	(a_4, b_4)

where a and b are nonnegative numbers such that

$$a + b = 1$$

Thus, the pair $\left[\dfrac{1}{4}, \dfrac{3}{4}\right]$ denotes the mixed strategies of using the first option one of four times, whereas the second one is used three out of four times. The pairs $[1, 0]$ and $[0, 1]$ are really pure strategies. Each calls for the consistent use of one option to the exclusion of the second one. Ruth's and Charlie's general mixed strategies are again denoted by $[1 - p, p]$ and $[1 - q, q]$, respectively. It proves convenient to refer to such a pair of strategies as a *mixed strategy pair*. When Ruth and Charlie use these mixed strategies in a nonzero-sum game, the expected payoff pair is computed in much the same manner as it was for the zero-sum games, except that each player focuses on his or her own payoffs. When the mixed strategies pair is $([1 - p, p], [1 - q, q])$, Ruth's expected payoff is denoted by $e_R(p, q)$ and Charlie's expected payoff is denoted by $e_C(p, q)$.

EXAMPLE 4.4.1. *Compute the expected payoffs when Ruth and Charlie use the mixed strategy pair ([0.3, 0.7], [0.6, 0.4]) in the game*

(1,2)	(3,−2)
(−1,0)	(0,4)

The auxiliary diagram

	0.6	0.4
0.3	(1,2)	(3,−2)
0.7	(−1,0)	(0,4)

yields

$$e_R(0.7, 0.4) = 0.3 \cdot 0.6 \cdot 1 + 0.3 \cdot 0.4 \cdot 3 + 0.7 \cdot 0.6 \cdot (-1) + 0.7 \cdot 0.4 \cdot 0$$
$$= 0.18 + 0.36 - 0.42 + 0 = 0.12$$

and

$$e_C(0.7, 0.4) = 0.3 \cdot 0.6 \cdot 2 + 0.3 \cdot 0.4 \cdot (-2) + 0.7 \cdot 0.6 \cdot 0 + 0.7 \cdot 0.4 \cdot 4$$
$$= 0.36 - 0.24 + 0 + 1.12 = 1.24$$

In 1950, John Nash (b. 1928) succeeded in extending the No Regrets aspect of pure equilibria to the mixed context. The significance of this achievement was underscored in 1994 when he was awarded the Nobel Prize for economics, in which discipline the concept of an equilibrium point has become an indispensable theoretical tool. The formulation of this new mixed version of the No Regrets principle is very similar to the pure one:

> After the repeated game is done, each player would like to rest assured that no other behavior on his part would have resulted in a better expected payoff.

A pair of mixed strategies is said to constitute a mixed Nash equilibrium of a mixed game if they result in payoffs such that neither player has cause to regret his or her choice of strategy. Nash's startling discovery was that unlike their pure kin, mixed Nash equilibria *always* exist. For the sake of completeness, Nash's theorem is stated in its full generality even though its terms cannot be fully defined here. This general statement is followed by a working restricted version.

NASH'S THEOREM 4.4.2. *Every n-person finite nonzero-sum game has an equilibrium point.*

The ensuing discussion and examples are confined mostly to two-person 2×2 nonzero-sum games. Our limited version of Nash's theorem can now be stated as follows:

THEOREM 4.4.3. *In any 2 × 2 nonzero-sum game, there is a mixed strategy pair* $[1 - \bar{p},$ $\bar{p}], [1 - \bar{q}, \bar{q}]$ *such that*

$$e_R(\bar{p}, \bar{q}) \geq e_R(p, \bar{q}) \text{ for all } p, \ 0 \leq p \leq 1 \tag{4.5}$$

and

$$e_C(\bar{p}, \bar{q}) \geq e_C(\bar{p}, q) \text{ for all } q, \ 0 \leq q \leq 1 \tag{4.6}$$

Inequality (4.5) of this theorem says that if both players adopt the stipulated strategies, then Ruth would have gained naught by adopting any other strategy. In other words, having used the recommended strategy $[1 - \bar{p}, \bar{p}]$, Ruth has no cause for regrets as long as Charlie sticks to $[1 - \bar{q}, \bar{q}]$. Inequality (4.6) makes a similar assertion about her opponent: Charlie has no cause to regret his employment of $[1 - \bar{q}, \bar{q}]$ as long as Ruth sticks to $[1 - \bar{p}, \bar{p}]$. The pair of strategies $([1 - \bar{p}, \bar{p}], [1 - \bar{q}, \bar{q}])$ described by this theorem constitute a *mixed Nash equilibrium*. The numbers $e_R(\bar{p}, \bar{q})$ and $e_C(\bar{p}, \bar{q})$ are the game's *mixed value pair*. Together, the mixed Nash equilibrium and the mixed value pair are a *Nash equilibrium solution* of the game.

All known proofs of Nash's theorem are nonconstructive. In other words, although they assert the existence of the mixed Nash equilibrium, they provide no method for actually deriving its constituent mixed strategies. The present chapter is confined to a description of a method for recognizing Nash equilibrium pairs as well as several applications of this concept to the real world. First, it is necessary to consider a preliminary issue.

Granted that in general neither player knows the opponent's strategy, still, *after* a game has been played, while trying to gauge whether she has cause for regrets, a player will be aware of the strategy that was used by her opponent. For this reason it is necessary at this point to consider the question of *optimal counterstrategies*. In other words, how should a player react if she knows her opponent's strategy? The guiding principle under these circumstances is given in the following theorem.

THEOREM 4.4.4. *If one player of a nonzero-sum game uses a fixed strategy, then the opponent has an optimal counterstrategy that is pure.*

EXAMPLE 4.4.5. *Suppose Ruth uses the strategy* $[0.2, 0.8]$ *in the game*

(3,2)	(2,1)
(0,3)	(4,4)

Find an optimal response for Charlie.

We know that Charlie has an optimal counterstrategy that is pure, and so we compute the expected payoffs that correspond to the two pure strategies that are available to him. The auxiliary diagrams below contain Charlie's payoffs only,

	1	0
0.2	2	1
0.8	3	4

	0	1
0.2	2	1
0.8	3	4

and they yield

$$0.2 \cdot 2 + 0.8 \cdot 3 = 2.8 \quad \text{for } [1, 0].$$
$$0.2 \cdot 1 + 0.8 \cdot 4 = 4.4 \quad \text{for } [0, 1].$$

Because Charlie is interested in maximizing his payoff, it follows that the pure strategy [0, 1], which yields him a payoff of 4.4, is an optimal response for him.

EXAMPLE 4.4.6. *Suppose Charlie uses the strategy [0.7, 0.3] in the game*

(3,2)	(2,1)
(0,3)	(4,4)

Find an optimal response for Ruth.

Because Ruth has an optimal counterstrategy that is pure, we compute the expected payoffs that correspond to the two pure strategies that are available to her. The auxiliary diagrams below contain Ruth's payoffs only

	0.7	0.3
1	3	2
0	0	4

	0.7	0.3
0	3	2
1	0	4

and they yield

$$0.7 \cdot 3 + 0.3 \cdot 2 = 2.7 \quad \text{for } [1, 0]$$
$$0.7 \cdot 0 + 0.3 \cdot 4 = 1.2 \quad \text{for } [0, 1]$$

Because Ruth is interested in maximizing her payoff, it follows that the pure strategy [1, 0], which yields her a payoff of 2.7, is an optimal response for her.

We now turn to the issue of recognizing mixed Nash equilibria. Suppose your mathematical consultant has provided you, at cost, with a mixed Nash equilibrium pair for a game that is of interest to you. How can you be sure that the mathematician has not made a mistake and that you are indeed getting your money's worth? Theorem 4.4.4 above provides us with a straightforward method for checking on the proposed strategy pair. One need merely verify that none of the pure strategies at each player's disposal provides that player with a better payoff than that resulting from the proposed Nash equilibrium pair.

EXAMPLE 4.4.7. *Verify that the mixed strategy pair* $([0.5, 0.5], [0.4, 0.6])$ *constitutes a mixed Nash equilibrium for the game*

(3,2)	(2,1)
(0,3)	(4,4)

Ruth's and Charlie's payoffs for the proposed Nash equilibrium pair are, respectively,

	0.4	0.6
0.5	(3,2)	(2,1)
0.5	(0,3)	(4,4)

$$e_R(0.5, 0.6) = 0.5 \cdot 0.4 \cdot 3 + 0.5 \cdot 0.6 \cdot 2 + 0.5 \cdot 0.4 \cdot 0 + 0.5 \cdot 0.6 \cdot 4$$
$$= 0.6 + 0.6 + 1.2 = 2.4$$

and

$$e_C(0.5, 0.6) = 0.5 \cdot 0.4 \cdot 2 + 0.5 \cdot 0.6 \cdot 1 + 0.5 \cdot 0.4 \cdot 3 + 0.5 \cdot 0.6 \cdot 4$$
$$= 0.4 + 0.3 + 0.6 + 1.2 = 2.5$$

Is it possible for Ruth to improve her payoff by switching to an alternate strategy? If so, then there must be a pure alternate strategy that allows her to accomplish this. However, the auxiliary diagrams below both yield payoffs of

	0.4	0.6
1	(3,2)	(2,1)
0	(0,3)	(4,4)

	0.4	0.6
0	(3,2)	(2,1)
1	(0,3)	(4,4)

$$0.4 \cdot 3 + 0.6 \cdot 2 = 2.4 \quad \text{for } [1, 0]$$
$$0.4 \cdot 0 + 0.6 \cdot 4 = 2.4 \quad \text{for } [0, 1]$$

both of which equal the value of $e_R(0.5, 0.6)$ computed above. This demonstrates that Ruth stands to gain nothing by switching to any alternate pure (or mixed) strategy. Thus, the proposed Nash equilibrium pair can cause Ruth no regrets.

Does Charlie have any cause for regrets? The auxiliary diagrams below yield payoffs

	1	0
0.5	(3,2)	(2,1)
0.5	(0,3)	(4,4)

	0	1
0.5	(3,2)	(2,1)
0.5	(0,3)	(4,4)

$$0.5 \cdot 2 + 0.5 \cdot 3 = 2.5 \quad \text{for } [1, 0]$$
$$0.5 \cdot 1 + 0.5 \cdot 4 = 2.5 \quad \text{for } [0, 1]$$

both of which equal the value of $e_C(0.5, 0.6)$ computed above. Thus, Charlie has no cause for regrets either.

The conclusion is that the strategy pair ([0.5, 0.5], [0.4, 0.6]) constitutes a Nash equilibrium pair for the given game.

The same method can also be used to recognize strategy pairs that are not Nash equilibria.

EXAMPLE 4.4.8. *Decide whether* $\left(\begin{bmatrix} 1 & 2 \\ 3 & 3 \end{bmatrix}, \begin{bmatrix} 1 & 5 \\ 6 & 6 \end{bmatrix} \right)$ *constitutes a Nash equilibrium pair for the game*

(5,1)	(0,0)
(0,0)	(1,5)

The expected payoffs corresponding to the proposed strategy pair are

	1/6	5/6
1/3	(5,1)	(0,0)
2/3	(0,0)	(1,5)

$$e_R\left(\frac{2}{3}, \frac{5}{6}\right) = \frac{1}{3} \cdot \frac{1}{6} \cdot 5 + \frac{2}{3} \cdot \frac{5}{6} \cdot 1 = \frac{5+10}{18} = \frac{15}{18} = \frac{5}{6}$$

$$e_C\left(\frac{2}{3}, \frac{5}{6}\right) = \frac{1}{3} \cdot \frac{1}{6} \cdot 1 + \frac{2}{3} \cdot \frac{5}{6} \cdot 5 = \frac{1+50}{18} = \frac{51}{18} = \frac{17}{6}$$

To see whether Ruth has any cause for regrets, we compute the payoffs of the auxiliary diagrams below as

	1/6	5/6
1	(5,1)	(0,0)
0	(0,0)	(1,5)

	1/6	5/6
0	(5,1)	(0,0)
1	(0,0)	(1,5)

$$\frac{1}{6} \cdot 5 = \frac{5}{6} \quad \text{for } [1, 0]$$

$$\frac{5}{6} \cdot 1 = \frac{5}{6} \quad \text{for } [0, 1]$$

Because these alternate payoffs equal the value of $e_R\left(\dfrac{2}{3},\dfrac{5}{6}\right)$, Ruth stands to gain nothing by relinquishing the given strategy $\begin{bmatrix} \dfrac{1}{3} & \dfrac{2}{3} \end{bmatrix}$. Passing on to Charlie, the auxiliary diagrams for the alternate pure strategies yield payoffs

	1	0
1/3	(5,1)	(0,0)
2/3	(0,0)	(1,5)

	0	1
1/3	(5,1)	(0,0)
2/3	(0,0)	(1,5)

$$\frac{1}{3}\cdot 1 = \frac{1}{3} \quad \text{for } [1, 0]$$

$$\frac{2}{3}\cdot 5 = \frac{10}{3} \quad \text{for } [0, 1]$$

Because the second of these, $\dfrac{10}{3} = 4.\overline{3}$, exceeds the value $\dfrac{17}{6} = 2.1\overline{6}$ of $e_C\left(\dfrac{2}{3},\dfrac{5}{6}\right)$, it follows that Charlie could improve his payoff by abandoning the proposed strategy of $\begin{bmatrix} \dfrac{1}{6} & \dfrac{5}{6} \end{bmatrix}$ in favor of the pure strategy $[0, 1]$. Hence, the given mixed strategy pair does not constitute a Nash equilibrium.

We offer the reader an alternate way of visualizing the defining properties of the Nash equilibrium. Suppose the values of $e_R(p, q)$ and $e_C(p, q)$, $p = 0, 0.2, 0.4, 0.6, 0.8, 1, q = 0, 0.2, 0.4, 0.6, 0.8, 1$ are tabulated for the game

(3,1)	(1,5)
(1,2)	(4,1)

in the form

p \ q	0	0.2	0.4	0.6	0.8	1
0	(3, 1)	(2.6, 1.8)	(2.2, 2.6)	(1.8, 3.4)	(1.4, 4.2)	(1, 5)
0.2	(2.6, 1.2)	(2.4, 1.8)	(2.2, 2.4)	(2, 3)	(1.8, 3.6)	(1.6, 4.2)
0.4	(2.2, 1.4)	(2.2, 1.8)	(2.2, 2.2)	(2.2, 2.6)	(2.2, 3)	(2.2, 3.4)
0.6	(1.8, 1.6)	(2, 1.8)	(2.2, 2)	(2.4, 2.2)	(2.6, 2.4)	(2.8, 2.6)
0.8	(1.4, 1.8)	(1.8, 1.8)	(2.2, 1.8)	(2.6, 1.8)	(3, 1.8)	(3.4, 1.8)
1	(1, 2)	(1.6, 1.8)	(2.2, 1.6)	(2.8, 1.4)	(3.4, 1.2)	(4,1)

Observe that in the $q = 0.4$ column, the first entry of all payoff pairs has the constant value of 2.2, and that in the $p = 0.8$ row, the second entry has the constant value of 1.8. Consequently, the mixed strategy pair ([0.2, 0.8], [0.6, 0.4]) is a Nash equilibrium because neither player stands to gain anything by changing his or her mixed strategy.

Although such a table can be a useful pedagogical tool, its construction is rather laborious and the details are greatly dependent on the exact numerical values of the constituent mixed strategies of the equilibrium. No more space therefore is devoted to such tables in this text.

The next two examples are meant to demonstrate the wide range of applicability of the notion of a mixed Nash equilibrium. They come from the disciplines of economics and biology, respectively.

4.4.1   The Job Applicants

In the previous section we modeled a situation wherein Ruth and Charlie were allowed to apply to one of two positions offering salaries $2a$ and $2b$, respectively, as the game

Charlie applies to

		Firm 1	Firm 2
Ruth applies to	Firm 1	(a, a)	$(2a, 2b)$
	Firm 2	$(2b, 2a)$	(b, b)

It was noted above that this game always has pure Nash equilibria whose exact values depend on the relative sizes of a and b. Specifically, if neither salary exceeds double the other, i.e., if

$$b \leq 2a \quad \text{and} \quad a \leq 2b$$

then the outcomes corresponding to the payoffs $(2a, 2b)$ and $(2b, 2a)$ constitute pure Nash equilibria. In that case, however, the game also possesses a mixed Nash equilibrium (see exercise 18) each of whose strategies is

$$\left[\frac{2a-b}{a+b}, \frac{2b-a}{a+b} \right]$$

The components of this common strategy can be interpreted as this model's predictions for the probabilities of Ruth (or Charlie) applying to the corresponding firm, i.e., if $a \leq 2b$ and $b \leq 2a$, then

$$\frac{2a-b}{a+b}$$ is the predicted probability of Ruth applying to firm 1

$$\frac{2b-a}{a+b}$$ is the predicted probability of Ruth applying to firm 2

For example, if firms 1 and 2 offer salaries of $20,000 and $30,000, respectively, then $a = \$10,000$, $b = \$15,000$, and this model predicts that the probability of a player applying for firm 2's position is

$$\frac{2 \cdot 15,000 - 10,000}{10,000 + 15,000} = \frac{20,000}{25,000} = 0.8$$

On the other hand, if either $a > 2b$ or $b > 2a$, or, in other words, if one of the salaries is more than double the other, then all the Nash equilibria are the pure ones that were already discussed in the previous section.

This analysis provides economists with a starting point for an investigation of the question of what effect wage differentials have on the pool of applicants. The details of this investigation fall outside the scope of this book.

4.4.2   An Evolutionary Game

In recent years, biological evolution has offered some applications of Nash equilibria as well. In many species, mating is preceded by a duel between the males. Stags shove each other with their antlers, and snakes have wrestling matches. The winner gets to mate. Similar intraspecies contests result from territorial disagreements. A surprising aspect of these contests is that nature seems to be pulling its punches. Some individuals run rather than fight. Stags do not gore each other's vulnerable sides, and in some species snakes do not bite each other. This is, of course, quite sensible behavior for the species, and the biologists J. Maynard Smith and G. A. Parker have constructed a game theoretic model for these contests in which this moderated ferocity is explained as a Nash equilibrium. An unusual feature of this game is that it pits the species in question against itself.

To define the game, we stipulate a species whose individuals engage in intra-species duels. Each such confrontation constitutes a play of the game. The individual members of the species are classified as either *hawks* or *doves*. A hawk always attacks in a confrontation, and a dove always runs away. The winner of a confrontation gets to mate or gets the better territory, and so he is in a better position to propagate his genes. In the event of an actual physical struggle, the loser sustains an injury. The payoff to the species of a single play consists of the effect the duel has on the individual's ability to reproduce, an elusive quantity called his *fitness*. It is assumed that the confrontation's winner's fitness is augmented by amount $2a$ and that a fight's loser's fitness is reduced by amount $2b$ (the 2 is introduced here into the payoffs to simplify the subsequent calculations). The precise payoffs are computed as follows.

Hawk vs. hawk: The two contestants continue fighting until one is injured. Inasmuch as each has a 50% chance of winning (and gaining $2a$) or losing (and losing $2b$), each of them is assigned the payoff

$$50\% \cdot 2a + 50\% \times (-2b) = a - b$$

Hawk vs. dove: Because the dove runs away, no physical struggle takes place; hawk gains $2a$ and dove neither gains nor loses anything.

Dove vs. dove: Again each contestant has a 50% chance of adding $2a$ to his fitness, but as there is no physical fight, there is no question of injury. Thus, each player is assigned the payoff

$$50\% \cdot 2a = a$$

The resulting nonzero-sum game has the array

	Hawk	Dove
Hawk	$(a - b,\ a - b)$	$(2a,\ 0)$
Dove	$(0,\ 2a)$	$(a,\ a)$

(4.7)

In this game both Ruth and Charlie represent the same species, and each play consists of a confrontation between some individuals. The pure strategy [1, 0] calls for the species to evolve hawks only. The pure strategy [0, 1] calls for the species to evolve doves only. A mixed strategy consists of evolving a mixture of both hawks and doves within the species. Such a strategy would presumably be encoded into the species' genes. However, mutations do occur with high frequency, and it is reasonable to assume that whatever strategy is now in effect could not be improved upon by any mutations. After all, if a change (that is internal to the species) could improve on its overall fitness, then, given the amount of time most species have been in existence, such a change would have in all likelihood taken place a long time ago. Thus, it is reasonable to conclude that the current ratio of hawks to doves in this species is a stable quantity that maximizes its overall fitness. It is now argued that the optimality of this ratio implies that it actually comes from a mixed Nash equilibrium of the nonzero-sum game (4.7).

Let us consider the mathematical analog of this optimality argument. Suppose the species is currently using the mixed strategy $[1 - \bar{p}, \bar{p}]$, i.e., its hawks-to-doves ratio is now $(1 - \bar{p}): \bar{p}$. Each individual's confrontation then augments his fitness by the expected payoff computed from the auxiliary diagram

	$1 - \bar{p}$	$\bar{p}$
$1 - \bar{p}$	$(a - b,\ a - b)$	$(2a,\ 0)$
$\bar{p}$	$(0,\ 2a)$	$(a,\ a)$

This value is

$$e_R(\bar{p}, \bar{p}) = e_C(\bar{p}, \bar{p}) = (1 - \bar{p})^2(a - b) + \bar{p}(1 - \bar{p})2a + \bar{p}^2 a$$

$$= \bar{p}^2(a - b - 2a + a) + \bar{p}(-2(a - b) + 2a) + 1(a - b)$$

$$= -\bar{p}^2 b + 2\bar{p}b + a - b$$

The reason $e_R(p, p)$ and $e_C(p, p)$ are equal is that both Ruth and Charlie represent the *same* species. The actual common value of these two expressions is in fact immaterial at this point. It is only necessary to keep in mind that $e_R(\bar{p}, \bar{p}) = e_C(\bar{p}, \bar{p})$.

Suppose now that a comparatively small subpopulation of this species mutates and begins breeding hawks and doves with a mixed strategy of $[1 - p, p]$ for some $p \neq p$. Because this subpopulation is presumed small, its individuals are mostly confronting normal individuals, and hence the expected increment to the mutated individual's fitness from each battle is the common value of $e_R(p, \bar{p})$ and $e_C(\bar{p}, p)$ that correspond to the auxiliary diagrams

		Normal population	
		$1 - \bar{p}$	$\bar{p}$
Mutated	$1 - p$	$(a - b, a - b)$	$(2a, 0)$
population	p	$(0, 2a)$	(a, a)

or

		Mutated population	
		$1 - p$	p
Normal	$1 - \bar{p}$	$(a - b, a - b)$	$(2a, 0)$
population	$\bar{p}$	$(0, 2a)$	(a, a)

The normal individual, however, still confronts mostly other normal individuals, and hence his fitness is augmented by the same quantity as before, namely,

$$e_R(\bar{p}, \bar{p}) = e_C(\bar{p}, \bar{p})$$

The above Panglossian argument that mutated strategies do not improve on the existing strategy entails, among others, the inequalities

$$e_R(\bar{p}, \bar{p}) \geq e_R(p, \bar{p}) \quad \text{and} \quad e_C(\bar{p}, \bar{p}) \geq e_C(\bar{p}, p)$$

These inequalities, however, are tantamount to the No Regrets guideline, because if we interpret game (4.7) as a game between the distinct opponents Ruth and Charlie, these inequalities say that neither Ruth nor Charlie could have improved their expected payoffs by changing from the strategy $[1 - \bar{p}, \bar{p}]$ to any other strategy $[1 - p, p]$. Thus, the existing hawks and doves breeding strategy $[1 - \bar{p}, \bar{p}]$ constitutes a mixed Nash equilibrium strategy of game (4.7). Let us examine the Nash equilibria of this game.

If $a > b$, then $a - b > 0$ and the outcome with payoff $(a - b, a - b)$ constitutes the only (pure or mixed) Nash equilibrium (see exercise 35). The associated pure strategy is $[1, 0]$. This can be interpreted as saying that if the advantage of winning outweighs the injuries that accompany a loss, then the species will breed only hawks.

If $a < b$, then (see exercise 36) there are two pure Nash equilibria with respective payoffs $(2a, 0)$ and $(0, 2a)$ and another mixed Nash equilibrium with strategies

$$[1 - \bar{p}, \bar{p}] = [1 - \bar{q}, \bar{q}] = \left[\frac{a}{b}, 1 - \frac{a}{b} \right]$$

The pure Nash equilibria can be ignored because each has $[1, 0]$ as the strategy for one player and $[0, 1]$ as the other player's strategy—an impossible situation given that both players stand for one and the same species. This leaves us with the mixed Nash equilibrium as the one that describes the species' hawks-to-doves ratio. Note that as a diminishes in comparison with b, this model predicts that the species will breed fewer and fewer hawks. In particular, if $a = 1$ and $b = 3$, i.e., if the effect of an injury from a duel outweighs the benefits of winning by a factor of 3 to 1, then this model predicts that only

$$\frac{a}{b} = \frac{1}{3}$$

of the individuals of the species will be hawks.

Given the inherently inaccurate nature of the payoffs, the final case, $a = b$, because of the exact equality it demands, is of course unlikely to occur in reality. Exercise 37 asserts that in this case hawks alone are bred.

An alternative interpretation of the Nash equilibrium strategy $[1 - \bar{p}, \bar{p}]$ in this evolutionary game is that it constitutes a ferocity index. On a scale of 0 (hawk) to 1 (dove), the quantity s denotes that species' willingness to fight that evolution has encoded into its genes. This could explain why, when fighting each other, the aforementioned snakes wrestle rather than bite.

4.4.3 $m \times n$ Nonzero-Sum Games

All the concepts developed here for 2×2 nonzero-sum games apply to larger games as well. An $m \times n$ nonzero-sum game is a rectangular array of m rows and n columns in which each entry is a pair of numbers. The entry in the ith row and jth column is denoted by $(a_{i,j}, b_{i,j})$. Given two mixed strategy pair $([p_1, p_2, ..., p_m], [q_1, q_2, ..., q_n])$, Ruth and Charlie's expected payoffs, denoted by e_R and e_C, respectively, are computed as

$$e_R = \text{sum of all } p_i \cdot q_j \cdot a_{i,j} \quad i = 1, 2, ..., m, j = 1, 2, ..., n$$
$$e_C = \text{sum of all } p_i \cdot q_j \cdot b_{i,j} \quad i = 1, 2, ..., m, j = 1, 2, ..., n$$

EXAMPLE 4.4.9. *Compute the expected payoffs when Ruth and Charlie use the mixed strategy pair ([0.5, 0.2, 0.3], [0.1, 0.2, 0.3, 0.4]) in the nonzero-sum game*

(1,2)	(0,−1)	(3,1)	(−2,0)
(1,−3)	(0,0)	(2,1)	(−1,1)
(3,2)	(1,1)	(−1,1)	(3,−1)

The auxiliary diagram

	0.1	0.2	0.3	0.4
0.5	(1,2)	(0,−1)	(3,1)	(−2,0)
0.2	(1,−3)	(0,0)	(2,1)	(−1,1)
0.3	(3,2)	(1,1)	(−1,1)	(3,−1)

yields the expected payoffs

$$
\begin{aligned}
e_R = {} & 0.5 \cdot 0.1 \cdot 1 + 0.5 \cdot 0.2 \cdot 0 + 0.5 \cdot 0.3 \cdot 3 + 0.5 \cdot 0.4 \cdot (-2) + 0.2 \cdot 0.1 \cdot 1 + \\
& 0.2 \cdot 0.2 \cdot 0 + 0.2 \cdot 0.3 \cdot 2 + 0.2 \cdot 0.4 \cdot (-1) + 0.3 \cdot 0.1 \cdot 3 + 0.3 \cdot 0.2 \cdot 1 + \\
& 0.3 \cdot 0.3 \cdot (-1) + 0.3 \cdot 0.4 \cdot 3 \\
= {} & 0.05 + 0.45 - 0.4 + 0.02 + 0.12 - 0.08 + 0.09 + 0.06 - 0.09 + 0.36 = 0.58
\end{aligned}
$$

$$
\begin{aligned}
e_C = {} & 0.5 \cdot 0.1 \cdot 2 + 0.5 \cdot 0.2 \cdot (-1) + 0.5 \cdot 0.3 \cdot 1 + 0.5 \cdot 0.4 \cdot 0 + 0.2 \cdot 0.1 \cdot (-3) + \\
& 0.2 \cdot 0.2 \cdot 0 + 0.2 \cdot 0.3 \cdot 1 + 0.2 \cdot 0.4 \cdot 1 + 0.3 \cdot 0.1 \cdot 2 + 0.3 \cdot 0.2 \cdot 1 + \\
& 0.3 \cdot 0.3 \cdot 1 + 0.3 \cdot 0.4 \cdot (-1) \\
= {} & 0.1 - 0.1 + 0.15 - 0.06 + 0.06 + 0.08 + 0.06 + 0.06 + 0.09 - 0.12 = 0.32
\end{aligned}
$$

Theorem 4.4.4 offers the same guidance in the search for optimality in this wider context as well.

EXAMPLE 4.4.10. *If Ruth uses the strategy [0.5, 0.2, 0.3] in the game of Example 4.4.9, find an optimal counterstrategy for Charlie.*

The auxiliary diagrams

	1	0	0	0
0.5	(1,2)	(0,−1)	(3,1)	(−2,0)
0.2	(1,−3)	(0,0)	(2,1)	(−1,1)
0.3	(3,2)	(1,1)	(−1,1)	(3,−1)

	0	1	0	0
0.5	(1,2)	(0,−1)	(3,1)	(−2,0)
0.2	(1,−3)	(0,0)	(2,1)	(−1,1)
0.3	(3,2)	(1,1)	(−1,1)	(3,−1)

	0	0	1	0
0.5	(1,2)	(0,−1)	(3,1)	(−2,0)
0.2	(1,−3)	(0,0)	(2,1)	(−1,1)
0.3	(3,2)	(1,1)	(−1,1)	(3,−1)

	0	0	0	1
0.5	(1,2)	(0,−1)	(3,1)	(−2,0)
0.2	(1,−3)	(0,0)	(2,1)	(−1,1)
0.3	(3,2)	(1,1)	(−1,1)	(3,−1)

yield the following payoffs for Charlie:

$$e_C = 0.5 \cdot 2 + 0.2 \cdot (-3) + 0.3 \cdot 2 = 1 - 0.6 + 0.6 = 1 \quad \text{for } [1, 0, 0, 0]$$
$$e_C = 0.5 \cdot (-1) + 0.2 \cdot 0 + 0.3 \cdot 1 = -0.5 + 0.3 = -0.2 \quad \text{for } [0, 1, 0, 0]$$
$$e_C = 0.5 \cdot 1 + 0.2 \cdot 1 + 0.3 \cdot 1 = 0.5 + 0.2 + 0.3 = 1 \quad \text{for } [0, 0, 1, 0]$$
$$e_C = 0.5 \cdot 0 + 0.2 \cdot 1 + 0.3 \cdot (-1) = 0.2 - 0.3 = -0.1 \quad \text{for } [0, 0, 0, 1]$$

Because 1 is the largest of these payoffs, it follows that both [1, 0, 0, 0] and [0, 0, 1, 0] are optimal counterstrategies for Charlie.

The procedure used to check on proposed Nash equilibrium pairs for 2×2 games can be applied in the more general context of $m \times n$ games as well.

EXAMPLE 4.4.11. *Show that ([0.4, 0.4, 0.2], [0.2, 0.6, 0, 0.2]) is a Nash equilibrium for the nonzero-sum game*

(0,1)	(1,0)	(1,0)	(0,1)
(1,0)	(0,1)	(1,0)	(2,−1)
(1,0)	(1,0)	(0,1)	(−1,2)

The auxiliary diagram

	0.2	0.6	0	0.2
0.4	(0,1)	(1,0)	(1,0)	(0,1)
0.4	(1,0)	(0,1)	(1,0)	(2,−1)
0.2	(1,0)	(1,0)	(1,0)	(−1,2)

yields the expected payoffs

$$e_R = 0.4 \cdot 0.6 \cdot 1 + 0.4 \cdot 0.2 \cdot 1 + 0.4 \cdot 0.2 \cdot 2 + 0.2 \cdot 0.2 \cdot 1 + 0.2 \cdot 0.6 \cdot 1 +$$
$$0.2 \cdot 0.2 \cdot (-1)$$
$$= 0.24 + 0.08 + 0.16 + 0.04 + 0.12 - 0.04 = 0.6$$

$$e_C = 0.4 \cdot 0.2 \cdot 1 + 0.4 \cdot 0.2 \cdot 1 + 0.4 \cdot 0.6 \cdot 1 + 0.4 \cdot 0.2 \cdot (-1) + 0.2 \cdot 0.2 \cdot 2$$
$$= 0.08 + 0.08 + 0.24 - 0.08 + 0.08 = 0.4$$

On the other hand, if Charlie sticks to his [0.2, 0.6, 0, 0.2] and Ruth experiments with her pure strategies, she gets the payoffs

$$
\begin{array}{ll}
0.6 \cdot 1 = 0.6 & \text{for } [1, 0, 0] \\
0.2 \cdot 1 + 0.2 \cdot 2 = 0.6 & \text{for } [0, 1, 0] \\
0.2 \cdot 1 + 0.6 \cdot 1 + 0.2 \cdot (-1) = 0.6 & \text{for } [0, 0, 1]
\end{array}
$$

all of which equal the value of e_R. Thus, Ruth has no reason to abandon her given strategy. If Ruth sticks to her strategy of [0.4, 0.4, 0.2] and Charlie experiments with his pure strategies, he gets the payoffs

$$
\begin{array}{ll}
0.4 \cdot 1 = 0.4 & \text{for } [1, 0, 0, 0] \\
0.4 \cdot 1 = 0.4 & \text{for } [0, 1, 0, 0] \\
0 & \text{for } [0, 0, 1, 0] \\
0.4 \cdot 1 + 0.4 \cdot (-1) + 0.2 \cdot 2 = 0.4 & \text{for } [1, 0, 0, 0]
\end{array}
$$

none of which is better than the value of e_C (note that one is actually worse). Thus, Charlie has no reason to regret staying with the given strategy either, and so the given strategy pair is indeed a Nash equilibrium.

EXERCISES 4.4

Compute the expected payoff when the games of Exercises 1–6 are played with the specified mixed strategy pairs.

1.

(4, 4)	(0, 1)
(1, 0)	(1, 1)

([0.7, 0.3], [0.2, 0.8])

2.

(4, 1)	(0, 2)
(1, 1)	(1, 0)

([0.2, 0.8], [0.5, 0.5])

3.

(4, 0)	(0, 1)
(1, 2)	(1, 5)

([0.3, 0.7], [0.8, 0.2])

4.

(4, 3)	(0, 2)
(1, 0)	(1, 1)

([0.7, 0.3], [0, 1])

5.

(4, 2)	(0, 1)
(1, 2)	(1, 0)

([1, 0], [0, 1])

(1,4)	(1,1)
(2,0)	(0,1)

6.   ([0.1, 0.9], [0.5, 0.5])

For each of the games in Exercises 7–12,

a)   Determine Ruth's optimal counterstrategy if Charlie uses the given strategy **C** and she knows it.

b)   Determine Charlie's optimal counterstrategy if Ruth uses the given strategy **R** and he knows it.

(3,2)	(2,4)
(2,3)	(4,−3)

7. $R = \begin{bmatrix} \dfrac{1}{2}, \dfrac{1}{2} \end{bmatrix}, \quad C = \begin{bmatrix} \dfrac{2}{3}, \dfrac{1}{3} \end{bmatrix}$

(3,2)	(2,4)
(2,3)	(4,−3)

8. $R = \begin{bmatrix} \dfrac{3}{4}, \dfrac{1}{4} \end{bmatrix}, \quad C = \begin{bmatrix} \dfrac{1}{2}, \dfrac{1}{2} \end{bmatrix}$

(3,2)	(2,4)
(2,3)	(4,−3)

9. $R = \begin{bmatrix} \dfrac{3}{4}, \dfrac{1}{4} \end{bmatrix}, \quad C = \begin{bmatrix} \dfrac{2}{3}, \dfrac{1}{3} \end{bmatrix}$

(2,−3)	(−1,3)	(−2,0)
(0,1)	(1,−2)	(0,0)
(−1,−2)	(2,3)	(3,−1)

10. $R = [0.2, 0.5, 0.3], \quad C = [0, 0.3, 0.7]$

(2,−3)	(−1,3)	(−2,0)	(4,−1)
(0,1)	(1,−2)	(0,0)	(−2,3)
(−1,−2)	(2,3)	(3,−1)	(2,2)

11. $R = [0.8, 0.1, 0.1],$
$C = [0.4, 0.3, 0.2, 0.1]$

(2,−3)	(−1,3)	(−2,0)
(0,1)	(1,−2)	(0,0)
(−1,−2)	(2,3)	(3,−1)
(4,−1)	(−2,3)	(2,2)

12. $R = [0.4, 0.4, 0.1, 0.1], \quad C = [0.4, 0.3, 0.3]$

For each of the games in Exercises 13–18, decide whether the mixed strategy pair constitutes a Nash equilibrium.

13.
(3,2)	(2,4)
(2,3)	(4,−3)

$$\left(\left[\frac{1}{2},\frac{1}{2}\right],\left[\frac{2}{3},\frac{1}{3}\right]\right)$$

14.
(3,2)	(2,4)
(2,3)	(4,−3)

$$\left(\left[\frac{3}{4},\frac{1}{4}\right],\left[\frac{1}{2},\frac{1}{2}\right]\right)$$

15.
(3,2)	(2,4)
(2,3)	(4,−3)

$$\left(\left[\frac{3}{4},\frac{1}{4}\right],\left[\frac{2}{3},\frac{1}{3}\right]\right)$$

16.
(2,−3)	(−1,3)
(0,1)	(1,−2)

$$\left(\left[\frac{1}{3},\frac{2}{3}\right],\left[\frac{1}{3},\frac{2}{3}\right]\right)$$

17.
(2,−3)	(−1,3)
(0,1)	(1,−2)

$$\left(\left[\frac{1}{3},\frac{2}{3}\right],\left[\frac{1}{2},\frac{1}{2}\right]\right)$$

18.
(2,−3)	(−1,3)
(0,1)	(1,−2)

$$\left(\left[\frac{3}{4},\frac{1}{4}\right],\left[\frac{1}{3},\frac{2}{3}\right]\right)$$

19. Decide whether the strategy pair $\left(\left[\frac{1}{8},0,\frac{5}{8},\frac{1}{4}\right],\left[\frac{1}{2},\frac{1}{2},0\right]\right)$ constitutes a Nash equilibrium for the following nonzero-sum game:

(1,1)	(0,2)	(−1,3)
(0,2)	(−1,3)	(2,0)
(1,1)	(0,2)	(1,1)
(−1,3)	(2,0)	(0,2)

20. Decide whether the strategy pair $\left(\left[\frac{1}{4},\frac{1}{4},0,\frac{1}{2}\right],\left[\frac{1}{2},0,\frac{1}{4},\frac{1}{4}\right]\right)$ constitutes a

Nash equilibrium for the following nonzero-sum game:

(1,0)	(0,1)	(1,0)	(−1,2)
(1,0)	(1,0)	(−1,2)	(1,0)
(−1,2)	(1,0)	(1,0)	(1,0)
(0,1)	(1,0)	(1,0)	(1,0)

21. Decide whether the strategy pair $\left(\left[0,\frac{2}{7},\frac{2}{7},\frac{3}{7}\right],\left[\frac{2}{7},\frac{3}{7},\frac{2}{7},0\right]\right)$ constitutes a Nash equilibrium for the following nonzero-sum game:

(−1,2)	(1,0)	(0,1)	(1,0)
(1,0)	(−1,2)	(1,0)	(0,1)
(0,1)	(1,0)	(−1,2)	(1,0)
(1,0)	(−1,2)	(1,0)	(−1,2)

22. Decide whether the strategy pair $\left(\left[\frac{2}{7},0,\frac{2}{7},\frac{3}{7}\right],\left[0,\frac{2}{7},\frac{3}{7},\frac{2}{7}\right]\right)$ constitutes a Nash equilibrium for the following nonzero-sum game:

(−1,2)	(1,0)	(0,1)	(1,0)
(1,0)	(−1,2)	(1,0)	(0,1)
(0,1)	(1,0)	(−1,2)	(1,0)
(1,0)	(−1,2)	(1,0)	(−1,2)

23. Decide whether the strategy pair $\left(\left[\frac{2}{7},0,\frac{2}{7},\frac{3}{7}\right],\left[\frac{2}{7},\frac{3}{7},\frac{2}{7},0\right]\right)$ constitutes a Nash equilibrium for the following nonzero-sum game:

(−1,2)	(1,0)	(0,1)	(1,0)
(1,0)	(−1,2)	(1,0)	(0,1)
(0,1)	(1,0)	(−1,2)	(1,0)
(1,0)	(−1,2)	(1,0)	(−1,2)

24. Show that the strategy pair $\left(\left[\dfrac{2a-b}{a+b}, \dfrac{2b-a}{a+b} \right], \left[\dfrac{2a-b}{a+b}, \dfrac{2b-a}{a+b} \right] \right)$ constitutes a mixed Nash equilibrium for the job applicants game whenever neither salary exceeds double the other.

25. Suppose $a = \$12{,}000$ and $b = \$25{,}000$ in the job applicants game. What is the predicted probability of Ruth applying to the position offered by

   a) Firm 1?   b) Firm 2?

26. Suppose $a = \$24{,}000$ and $b = \$20{,}000$ in the job applicants game. What is the predicted probability of Ruth applying to the position offered by

   a) Firm 1?   b) Firm 2?

27. Suppose $a = \$10{,}000$ and $b = \$12{,}000$ in the job applicants game. What is the predicted probability of Ruth applying to the position offered by

   a) Firm 1?   b) Firm 2?

28. Suppose $a = \$50{,}000$ and $b = \$20{,}000$ in the job applicants game. What is the predicted probability of Ruth applying to the position offered by

   a) Firm 1?   b) Firm 2?

29. Suppose $a = \$12{,}000$ and $b = \$18{,}000$ in the job applicants game. What is the predicted probability of Ruth applying to the position offered by

   a) Firm 1?   b) Firm 2?

30. Suppose $a - \$11{,}000$ and $b = \$26{,}000$ in the job applicants game. What is the predicted probability of Ruth applying to the position offered by

   a) Firm 1?   b) Firm 2?

31. Suppose $a = 2$ and $b = 3$ in the Maynard Smith & Price evolutionary game. What is the predicted proportion of

   a) Hawks?   b) Doves?

32. Suppose $a = 3$ and $b = 2$ in the Maynard Smith & Price evolutionary game. What is the predicted proportion of

   a) Hawks?   b) Doves?

33. Suppose $a = 15$ and $b = 9$ in the Maynard Smith & Price evolutionary game. What is the predicted proportion of

   a) Hawks?   b) Doves?

34. Suppose $a = 6$ and $b = 15$ in the Maynard Smith & Price evolutionary game. What is the predicted proportion of

   a) Hawks?   b) Doves?

35. Show that if $a > b$ in the evolutionary game, then there is only one pure Nash equilibrium and identify it.

36. Show that if $a < b$ in the evolutionary game, then there are two pure Nash equilibria and identify them. Also show that the two strategies $R = C = [a/b, a/b]$ constitute a mixed Nash equilibrium.

37. Show that if $a = b$, then the evolutionary game has only one pure Nash equilibrium and no other mixed Nash equilibria.

4.5 The Solution of Zero-Sum Games

This exposition of game theory began with a discussion of zero-sum games in which each player's gain is the other player's loss and vice versa. The "solution" of these games is now derived within the wider context of nonzero-sum games and their mixed equilibria. Historically, the solution of zero-sum games preceded Nash's theory by two decades. We decided to give this nonchronological account because it allows for a faster development of the subject matter.

As was noted in Section 4.2, payoffs of zero-sum games are usually denoted by a single number that simultaneously designates both Ruth's gain and Charlie's loss. Hence, the array of the typical 2×2 zero-sum game has the abbreviated form

Charlie

Ruth

a_1	a_2
a_3	a_4

(4.8)

as well as the full form

Charlie

Ruth

$(a_1, -a_1)$	$(a_2, -a_2)$
$(a_3, -a_3)$	$(a_4, -a_4)$

(4.9)

We reexamine the notions of pure maximin strategies and values, pure Nash equilibria, and mixed Nash equilibria in the context of zero-sum games.

4.5.1 Pure Maximin Strategies and Values

Because the entries in the array of a zero-sum game denote Ruth's payoffs, her *pure maximin strategy* and *value* remain the same, regardless of the context. On the other hand, the search for guarantees that lead to Charlie's maximin strategy for nonzero-sum games needs to be rephrased in terms of Ruth's payoffs. The logic of guarantees dictates that Charlie should select the column that gives him the best possible guarantee. Once Charlie has selected a column, Ruth cannot gain more than the maximum entry of that column. Consequently, this logic directs Charlie to select that column whose maximum entry is as small as possible. For that purpose we record at the bottom of each column of the zero-sum array that column's maximum element and then circle the smallest of these maxima. Because what we have here is the smallest of a set of maxima, this strategy is called the *pure minimax strategy* and the corresponding circled entry is the *pure minimax value*.

EXAMPLE 4.5.1. *Determine the pure maximin and minimax strategies and values of the zero-sum game with the array*

4	−2	2	−3	5
5	3	0	2	−1
6	3	2	3	1

To find Ruth's pure maximin strategy, record to the right of each row its minimum entry and then circle the largest of these (see below). Accordingly,

Row
min

4	−2	2	−3	5	−3
5	3	0	2	−1	−1
6	3	2	3	1	①

Column max   6   3   ②   3   5

This means that Ruth's maximin strategy is to play the third row, and this guarantees her a payoff of at least 1. To find Charlie's pure minimax strategy, record at the bottom of each column that column's maximum entry and then circle the smallest of these. Accordingly, this means that Charlie's minimax strategy is to play the third column, and his minimax value is 2. What these numbers mean is that by playing the third row, Ruth can guarantee a payoff of at least 1, and by playing the third column, Charlie can guarantee that her payoff will never exceed 2.

Charlie's pure maximin and minimax strategies are always identical. His pure maximin and minimax *values*, however, are the negatives of each other. The reader should also take note of the following fact:

In a zero-sum game, every column maximum is greater than or equal to every row minimum.   (4.10)

This observation will soon prove its usefulness. It is justified by an examination of the zero-sum game below wherein a is a typical payoff, x is a typical column maximum, and y is a typical row minimum. By definition,

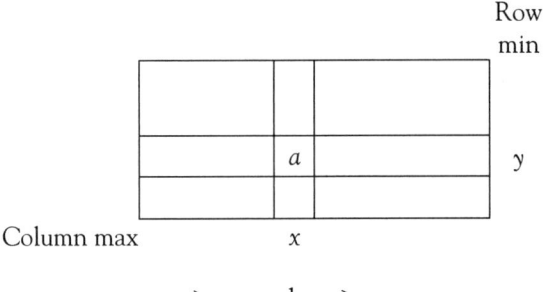

$$x \geq a \quad \text{and} \quad a \geq y$$

and so it follows that

$$x \geq y$$

where x and y denote *any* column maximum and *any* row minimum, respectively.

4.5.2 Pure Nash Equilibria of Zero-Sum Games

Suppose, for the sake of argument, that $(a_1, -a_1)$ is the payoff of a pure Nash equilibrium of the general games (4.8) and (4.9) above. This means that

$$a_1 \geq a_3 \quad \text{and} \quad -a_1 \geq -a_2$$

or, because multiplication by minus reverses inequalities,

$$a_1 \geq a_3 \quad \text{and} \quad a_1 \leq a_2$$

In other words, a_1 is the largest payoff in its column and the smallest in its row. These properties characterize the pure Nash equilibria of all zero-sum games, regardless of their dimensions. That is to say, the payoff a of the abbreviated zero-sum array below corresponds to a pure Nash equilibrium of the full array provided it has the following two properties:

1. a is a largest entry of its column.

2. a is a smallest entry of its row.

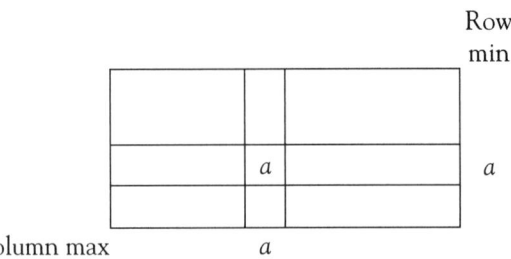

Such entries are called *saddle points*, and the circled entries of the two zero-sum games below are both such saddle points:

I

9	−2	−5
5	1	−9
3	②	5
−5	0	1

II

5	−2	1	−3	4
0	−2	5	0	−1
9	⊝1	0	2	1

They are, of course, pure Nash equilibria of the full array, but in zero-sum games they are also simultaneously Ruth's pure maximin and Charlie's pure minimax values. The reason for this lies in observation (4.10) above. Because every column maximum is at least as great as every row minimum, and because a is both a column maximum *and* a row minimum, it follows that it must be both the least column maximum and the greatest row minimum, i.e., a is both the pure minimax and the pure maximin values (see examples below). This observation also provides a method for finding the saddle points of zero-sum games when such points exist. List the row minima and the column maxima as was done for the maximin and minimax values.

I

9	−2	−5	−5
5	1	−9	−9
3	②	5	②
−5	0	1	−5
9	②	5	

II

5	−2	1	−3	4	−3
0	−2	5	0	−1	−2
9	⊝1	0	2	1	⊝1
9	⊝1	5	2	4	

If one of the row minima equals one of the column maxima, then the corresponding entry is a saddle point. If no such equality occurs, then the game does not have a saddle point. Zero-sum games that do have saddle points are said to be *strictly determined*.

4.5.3   The Battle of the Bismarck Sea

In March 1943 a battle took place near New Guinea that can be described as a strictly determined zero-sum 2×2 game. The players, so to speak, were the Allies (Ruth) and the Japanese (Charlie). The latter were consolidating their positions and found it necessary to transport troops from Rabaul to Lae, and the former were fore-warned of these plans by intelligence reports which indicated that a Japanese troop and supply convoy was assembling at Rabaul. This convoy was expected to unload at Lae and could reach it by traveling either north or south of New Britain, each route requiring 3 days. Rain and poor visibility were expected for the northern route, whereas clear weather and good visibility were forecast for south of the island.

The Allies also faced a choice. They could concentrate their reconnaissance on either the northern or the southern routes. It was estimated that if the Japanese chose the northern route, then it would take the allies either 1 or 2 days to locate the convoy, depending on whether the reconnaissance was concentrated on the northern or southern routes. On the other hand, if the Japanese chose the southern option, it would take either 1 day or no time at all to locate the convoy, again depending on the allies' decision as to where to concentrate the search. For the remainder of the days, the Japanese convoy would be exposed to allied aerial bombing.

This battle can be summarized as the following 2×2 zero-sum game whose payoffs denote the number of bombing days left for the allies:

		Japanese choice of route	
		North	South
Allied concentration of reconnaisance	North	2 days	2 days
	South	1 day	3 days

This game has a saddle point whose pure maximin and minimax strategies are both [1, 0], i.e., North. In the event, this is exactly what transpired. The Japanese convoy used the northern route and that was where the allies expected them. General George C. Kenney, the allied commander, offered the following description of decision process: "We decided to gamble that the Nip would take the northern route." The Japanese general's rationale for not gambling and trying to fool the Allies by taking the unexpected southern route is unknown. In all likelihood, he was trying to avoid the worst possible scenario wherein his forces would take the southern route and be exposed to 3 days of punishing bombardment. In other words, he was probably using the maximin logic.

4.5.4   Mixed Nash Equilibria of Zero-Sum Games

By the Nash Theorem 4.4.3, every game, zero-sum games included, has a mixed Nash equilibrium. Because the payoff pairs of the full arrays of zero-sum games add up to zero, the mixed value pairs of the equilibrium will also have a sum of zero. In other words, if Ruth's equilibrium payoff is e, then Charlie's equilibrium payoff is $-e$. This expected payoff e has a meaning that is similar to that of the saddle point payoff of strictly determined games.

Suppose this mixed Nash equilibrium consists of the mixed strategies $R = [1 - p, p]$ for Ruth and $C = [1 - \overline{q}, \overline{q}]$ for Charlie. Because of the No Regrets property of the Nash equilibrium, Charlie cannot improve his expected payoff by changing his mind.

Consequently, by playing R, Ruth guarantees herself a minimum expected payoff of e. Could there be another mixed strategy, say R′, which would guarantee Ruth some larger minimum expectation $e' > e$? The answer is no, and the reason is that by playing C, Charlie can guarantee himself a payoff of at least $-e$; in other words, the strategy C enables Charlie to hold Ruth's expectation down to e. Hence, e is also the largest minimum expectation that Ruth can guarantee. Because this guarantee is provided by strategy R, this strategy is called the mixed *maximin strategy*.

Similarly, again because of the No Regrets principle, Ruth cannot improve her expected payoff by changing her mind. In other words, by playing C, Charlie can guarantee that Ruth's expectation will be held down to a maximum of e. Could there be another mixed strategy, say C′, which would enable Charlie to hold Ruth's expectation down to some smaller value $e'' < e$? The answer is no, and the reason is that by playing R, Ruth can guarantee herself a payoff of at least e. Hence, e is the smallest maximum expectation that Charlie can hold Ruth down to. Because this is provided to Charlie by strategy C, this strategy is called the mixed *minimax strategy*. The expectation e is the *value* of the game, and this value together with the mixed maximin and minimax strategies constitute the *solution of the game*.

The following theorem is the central theorem of the theory of zero-sum games. Although this theorem is presented here as a consequence of Nash's equilibrium theorem, it was first stated and proved by von Neumann as early as 1928, thus preceding Nash's theorem by over 20 years. It did not attain wide recognition until the publication of von Neumann and Morgenstern's seminal book, *The Theory of Games and Economic Behavior*, in 1944. It is now commonly known as *the Minimax theorem*.

MINIMAX THEOREM. *For every zero-sum game there is a number v which has the following properties:*

> a) *Ruth has a mixed strategy that guarantees her an expected payoff of <u>at least</u> v.*
>
> b) *Charlie has a mixed strategy that guarantees that (Ruth's) expected payoff will be <u>at most</u> v.*

The nature of the guarantees is such that if both players use their recommended strategies, then the expected payoff will be exactly v. In other words, when Ruth uses the maximin strategy and Charlie uses the minimax strategy, Ruth can expect to win v units per play (on the average). For this reason, the number v is called the *value* of the game. The value of a game together with its maximin and minimax strategies constitute the *solution* of the game.

In contrast with Nash equilibria of nonzero-sum games, zero-sum 2×2 games are easily solved. The solution of strictly determined games was already described above. For any nonstrictly determined game

a_1	a_2
a_3	a_4

(4.11)

Ruth's oddments are defined to be one of the pairs

$$[a_4 - a_3, a_1 - a_2] \quad \text{or} \quad [a_3 - a_4, a_2 - a_1]$$

which consists of positive numbers alone (the game being nonstrictly determined, one of these pairs will have this property). For example, for game

2	3
4	1

(4.12)

Ruth's oddments are

$$[4 - 1, 3 - 2] = [3, 1]$$

whereas for the game

5	-2
1	4

(4.13)

Ruth's oddments are

$$[4 - 1, 5 - (-2)] = [3, 7]$$

The maximin strategy is obtained from the oddments when they are each divided by their sum. In game (4.12), the maximin strategy is

$$\left[\frac{3}{3+1}, \frac{1}{3+1} \right] = \left[\frac{3}{4}, \frac{1}{4} \right] = [0.75, 0.25]$$

and the oddments [3, 7] of game (4.13) yield the maximin strategy

$$\left[\frac{3}{3+7}, \frac{7}{3+7} \right] = \left[\frac{3}{10}, \frac{7}{10} \right] = [0.3, 0.7]$$

Charlie's oddments in game (4.11) consist of one of the pairs

$$[a_4 - a_2, a_1 - a_3] \quad \text{or} \quad [a_2 - a_4, a_3 - a_1]$$

which consists of two positive numbers. These oddments are converted to the minimax strategy in the same manner as above. Thus, for game (4.12) Charlie's oddments are

$$[3 - 1, 4 - 2] = [2, 2]$$

and his minimax strategy is

$$\left[\frac{2}{2+2}, \frac{2}{2+2} \right] = \left[\frac{2}{4}, \frac{2}{4} \right] = [0.5, 0.5]$$

For game (4.13) Charlie's oddments are

$$[4 - (-2), 5 - 1] = [6, 4]$$

and his minimax strategy is

$$\left[\frac{6}{6+4}, \frac{4}{6+4}\right] = \left[\frac{6}{10}, \frac{4}{10}\right] = [0.6, 0.4]$$

The value of the game is computed by means of the usual auxiliary diagram. Thus, for game (4.13) we have

	0.6	0.4
0.3	5	-2
0.7	1	4

giving a value of

$$v = 0.3 \cdot 0.6 \cdot 5 + 0.3 \cdot 0.4 \cdot (-2) + 0.7 \cdot 0.6 \cdot 1 + 0.7 \cdot 0.4 \cdot 4$$
$$= 0.9 - 0.24 + 0.42 + 1.12 = 2.2$$

4.5.5   Caveat

It is important to stress that the oddments method only applies to *nonstrictly deter-mined* games. If the method is brought to bear on a strictly determined game, the derived answer, although it may look legitimate, in general will be wrong. Thus, the general procedure when solving a zero-sum 2×2 game calls for first deciding whether or not it is strictly determined. If it is, the decision process of Section 4.5.2 yields the solution. If the game is not strictly determined, then the oddments method should be applied.

THE RIVER TALE   (J. D. Williams, 1986). *Steve is approached by a stranger who sug-gests that they match coins. Steve says that it's too hot for vigorous exercise. The stranger says, "Well then, let's just lie here and speak the words* heads *or* tails—*and to make it interesting I'll give you $30 when I call* tails *and you call* heads, *and $10 when it's the other way round. And—just to make it fair—you give me $20 when we match."*

The payoffs may look fair, but they are not. This is the zero-sum game

		Stranger	
		Heads	Tails
Steve	Heads	-20	30
	Tails	10	-20

Steve's oddments are

$$[10 - (-20), 30 - (-20)] = [30, 50]$$

and the stranger's are

$$[30 - (-20), 10 - (-20)] = [50, 30]$$

Hence, their respective mixed maximin and minimax strategies are [3/8, 5/8] and [5/8, 3/8]. The auxiliary diagram is

	5/8	3/8
3/8	-20	30
5/8	10	-20

yielding the value

$$v = \frac{3}{8} \cdot \frac{5}{8} \cdot (-20) + \frac{3}{8} \cdot \frac{3}{8} \cdot 30 + \frac{5}{8} \cdot \frac{5}{8} \cdot 10 + \frac{5}{8} \cdot \frac{3}{8} \cdot (-20) =$$

$$\frac{-300 + 270 + 250 - 300}{8 \cdot 8} = -\frac{80}{64} = -\$1.25$$

In other words, should Steve chose to play this game, he will lose, on the average, $1.25 each time he plays.

4.5.6   Bombing Sorties

This game, which was discussed in great detail in Section 4.1, can now be solved. If the immaterial % sign is dropped from the payoff, then this game has array

80	100
90	50

Ruth's oddments are

$$[90 - 50, 100 - 80] = [40, 20]$$

yielding a maximin strategy of

$$\left[\frac{40}{60}, \frac{20}{60} \right] = \left[\frac{2}{3}, \frac{1}{3} \right]$$

Charlie's oddments are

$$[100 - 50, 90 - 80] = [50, 10]$$

yielding a minimax strategy of

$$\left[\frac{50}{60}, \frac{10}{60}\right] = \left[\frac{5}{6}, \frac{1}{6}\right]$$

The auxiliary diagram is

	5/6	1/6
2/3	80	100
1/3	90	50

and so the game's value is

$$v = \frac{2}{3} \cdot \frac{5}{6} \cdot 80 + \frac{2}{3} \cdot \frac{1}{6} \cdot 100 + \frac{1}{3} \cdot \frac{5}{6} \cdot 90 + \frac{1}{3} \cdot \frac{1}{6} \cdot 50 =$$

$$\frac{800 + 200 + 450 + 50}{18} = \frac{1500}{18} = 83.\overline{3}\%$$

This solution is to be interpreted as follows. It is the game's theoretic recommendation that Ruth place the bomb on the support plane one out of three times (on the average) and that Charlie attack the support plane one out of six times (again, on the average). If they follow these recommendations, Ruth guarantees herself an expected bomb delivery rate of at least 83.$\overline{3}$%, whereas Charlie guarantees that Ruth's expectation will not exceed 83.$\overline{3}$%.

The next example describes a social zero-sum game whose array has more than two rows. Although the solution of such games falls outside the scope of this text, the validity of such solutions can be verified by the same means used in the previous section for verifying Nash equilibria. All that is needed is to simply replace each payoff a of the zero-sum game by the payoff pair $(a, -a)$ and then proceed with the verification process.

THE JAMAICAN FISHING VILLAGE. *The anthropologist W. C. A. Davenport observed the inhabitants of a certain Jamaican fishing village for 2 years in the early 1950s. These fishermen possessed 26 fishing canoes manned each by a captain and two or three crewmen. The fishing took the form of setting pots (traps) and drawing from them. The fishing grounds were divided into inside and outside banks. The inside banks lay from 5 to 15 miles offshore, whereas the outside banks lay beyond. The crucial factor that distinguished between the two areas was the occasional presence of very strong currents in the latter, which rendered fishing impossible. Accordingly, each captain must decide each day on a fishing strategy for that day. He could (1) set all his pots inside, (2) set all the pots outside, or (3) set some of the pots inside and some outside.*

Davenport modeled this situation as a 3 × 2 game whose players are the village and the environment. On any day, each of the 26 captains *decided on an option for the* village by adopting one of the three available fishing strategies for that day. The environment "decided" on its option by either sending a current or not. Based on his observations of the local marketplace and the costs accrued by the captains, Davenport estimated the payoffs of the various possibilities as follows:

Environment

		Current	No current
	Inside	17.3	11.5
Village	In-out	5.2	17.0
	Outside	−4.4	20.6

The monetary unit was the Pound, and the reason for the one negative payoff was that the Captain must pay his crew for each outing, regardless of the catch. Davenport went on to treat this as a zero-sum two-person game, solved it as such, and compared the game's maximin strategy with the actual distribution of the captains' choices.

The solution of this game is as follows: (see exercise 49):

Mixed maximin strategy: [0.67, 0.33, 0]

Mixed minimax strategy: [0.31, 0.69]

Value: 13.31

This is to be understood as saying that game theory's recommendation to the village is that 67% of its fishing should be done in the inside banks exclusively, 33% as an inside-outside combination, and none of the fishermen should dedicate themselves to fishing in the outside banks alone. This way the village can guarantee its fishermen an expected payoff of at least 13.31 per outing.

Davenport observed that 18 of the captains (69%) fished only in the inside banks, 8 (31%) adopted the inside-outside combination, and none restricted their fishing to the outside banks alone—a remarkable fit between theory and observation. On the other hand, during his 2-year stay, Davenport observed that a current was present in the outside bank on 25% of the days. This could be interpreted as saying that the environment was using the mixed strategy of [0.25, 0.75]. This is not as exciting a fit with the minimax strategy of [0.31, 0.69] as we had for the village's maximin strategy, but it is still close. Although this is not to be taken as evidence for the anthropomorphic view of nature, it should be pointed out that the game's payoffs are dependent on many factors, including the frequencies of the currents and the village's fishing strategies. Is it possible that this dependence might cause the game's entries to gravitate toward values whose minimax strategy agrees with the actual frequency of the current?

EXERCISES 4.5

Solve the games in Exercises 1–40.

1.
$$\begin{array}{|c|c|} \hline 1 & 3 \\ \hline 2 & 1 \\ \hline \end{array}$$

2.
$$\begin{array}{|c|c|} \hline 3 & 2 \\ \hline 0 & 2 \\ \hline \end{array}$$

3.
$$\begin{array}{|c|c|} \hline 2 & 1 \\ \hline 3 & 1 \\ \hline \end{array}$$

4.
$$\begin{array}{|c|c|} \hline 2 & 1 \\ \hline 1 & 3 \\ \hline \end{array}$$

5.
$$\begin{array}{|c|c|} \hline 2 & 1 \\ \hline 4 & 6 \\ \hline \end{array}$$

6.
$$\begin{array}{|c|c|} \hline 1 & -3 \\ \hline -2 & 1 \\ \hline \end{array}$$

7.
$$\begin{array}{|c|c|} \hline 1 & 3 \\ \hline -2 & 0 \\ \hline \end{array}$$

8.
$$\begin{array}{|c|c|} \hline 0 & -1 \\ \hline 0 & 1 \\ \hline \end{array}$$

9.
$$\begin{array}{|c|c|} \hline 2 & 4 \\ \hline 0 & -1 \\ \hline \end{array}$$

10.
$$\begin{array}{|c|c|} \hline 1 & 1 \\ \hline 2 & -1 \\ \hline \end{array}$$

11.
$$\begin{array}{|c|c|} \hline 1 & 1 \\ \hline -3 & 1 \\ \hline \end{array}$$

12.
$$\begin{array}{|c|c|} \hline -2 & 1 \\ \hline 0 & 3 \\ \hline \end{array}$$

13.
$$\begin{array}{|c|c|} \hline 5 & -1 \\ \hline -4 & 3 \\ \hline \end{array}$$

14.
$$\begin{array}{|c|c|} \hline -1 & 3 \\ \hline 6 & -4 \\ \hline \end{array}$$

15.
$$\begin{array}{|c|c|} \hline 8 & 2 \\ \hline 2 & -8 \\ \hline \end{array}$$

16.
$$\begin{array}{|c|c|} \hline 3 & 1 \\ \hline 1 & 2 \\ \hline \end{array}$$

17.
$$\begin{array}{|c|c|} \hline -3 & 1 \\ \hline -4 & -2 \\ \hline \end{array}$$

18.
$$\begin{array}{|c|c|} \hline 3 & -3 \\ \hline -6 & 1 \\ \hline \end{array}$$

19.
$$\begin{array}{|c|c|} \hline 0 & 0 \\ \hline 0 & -2 \\ \hline \end{array}$$

20.
$$\begin{array}{|c|c|} \hline -1 & -2 \\ \hline 1 & 3 \\ \hline \end{array}$$

21.
$$\begin{array}{|c|c|} \hline -1 & -2 \\ \hline 0 & -3 \\ \hline \end{array}$$

22.
$$\begin{array}{|c|c|} \hline 8 & -2 \\ \hline 2 & 8 \\ \hline \end{array}$$

23.
$$\begin{array}{|c|c|} \hline 2 & 4 \\ \hline 0 & -1 \\ \hline \end{array}$$

24.
$$\begin{array}{|c|c|} \hline 2 & -2 \\ \hline 2 & 2 \\ \hline \end{array}$$

25.
$$\begin{array}{|c|c|} \hline 3 & -1 \\ \hline 0 & 3 \\ \hline \end{array}$$

26.
$$\begin{array}{|c|c|} \hline 1 & 0 \\ \hline 2 & 0 \\ \hline \end{array}$$

27.
$$\begin{array}{|c|c|} \hline 1 & -2 \\ \hline 3 & 0 \\ \hline \end{array}$$

28.
$$\begin{array}{|c|c|} \hline 2 & 1 \\ \hline 2 & 4 \\ \hline \end{array}$$

29.
$$\begin{array}{|c|c|} \hline 0 & -1 \\ \hline -1 & 1 \\ \hline \end{array}$$

30.
$$\begin{array}{|c|c|} \hline -2 & 2 \\ \hline 5 & -5 \\ \hline \end{array}$$

33.

0	2
1	-1
2	3
-1	-2

31.

3	1	2
5	-1	3

32.

0	-2	7	2
3	-1	-1	2

34.

-5	5	0
2	-2	1
4	3	2

35.

1	-2	0	-3
-1	-2	0	1
5	-4	6	-3

36.

1	2	3
0	1	2
-1	0	1
-2	-1	0

37.

0	2	-4	0
-2	0	-2	4
4	2	0	6
0	0	-6	0

38.

-1	-2	0	1	0
3	4	2	2	4
2	0	1	-1	-3
1	-1	5	2	6

39.

2	1	1	3
6	-1	-7	8
-8	1	7	-6
5	0	-5	1
-1	-2	2	3

40.

5	2	-5	6	0
-2	0	2	0	-3
5	1	0	-1	-2
3	0	-1	2	-2
0	2	-1	0	-1

Describe the following situations as mathematical zero-sum games and solve them.

41.   *The Birthday* (J. D. Williams, 1986). Frank is hurrying home late, after a particularly grueling day, when it pops into his mind that today is Kitty's birthday! Or is it? Everything is closed except the florists. If it is not her birthday and he brings no gift, the situation will be neutral, i.e., payoff 0. If it is not and he comes in bursting with roses and obviously confused, he may be subjected to the Martini test, but he will emerge in a position of strong one-upness, which is worth 1. If it is her birthday and he has, clearly, remembered it, that is worth somewhat more, say 1.5. If he has forgotten it he is down like a stone, say, -10.

42. *The Hi-Fi* (J. D. Williams, 1986). The firm of Gunning & Kappler manufactures an amplifier. Its performance depends critically on the characteristics of one small inaccessible condenser. This normally costs Gunning & Kappler $1, but they are set back a total of $10, on the average, if the original condenser is defective. There are some alternatives open to them. It is possible for them to buy a superior quality condenser at $6, which is fully guaranteed; the manufacturer will make good the condenser and the costs incurred in getting the amplifier to operate. There is also available a condenser covered by an insurance policy that states, in effect, "If it is our fault, we will bear the costs and you get your money back." This item costs $10. (This is a 3×2 game that Gunning & Kappler is playing against Nature, whose options are to supply either a defective or a nondefective condenser.)

43. *The Huckster* (J. D. Williams, 1986). Merrill has a concession at the Yankee Stadium for the sale of sunglasses and umbrellas. The business places quite a strain on him, the weather being what it is. He has observed that he can sell about 500 umbrellas when it rains and about 100 when it shines; in the latter case he also can dispose of about 1000 sunglasses. Umbrellas cost him 50 cents and sell for $1 (this is 1954); glasses cost 20 cents and sell for 50 cents. He is willing to invest $250 in the project. Everything that isn't sold is a total loss (the children play with them).

44. *The Coal Problem* (J. D. Williams, 1986). On a sultry summer afternoon, Hans' wandering mind alights upon the winter coal problem. It takes about 15 tons to heat his house during a normal winter, but he has observed extremes when as little as 10 tons and as much as 20 were used. He also recalls that the price per ton seems to fluctuate with the weather, being $10, $15, and $20 a ton during mild, normal, and severe winters, respectively. He can buy now, however, at $10 a ton. He considers three pure strategies, namely, to buy 10, 15, or 20 tons now and the rest, if any, later. He will be moving to California in the spring and he cannot take excess coal with him.

45. The payoffs in the game below denote dollars. Ruth and Charlie are about to play this game 100 times.

3	−1
1	2

a) Does Ruth have a strategy that guarantees her winning something each time the game is played?

b) Does Ruth have a strategy that guarantees her winning a total of $80 or more?

c) Does Ruth have a strategy that guarantees her winning a total of $120 or more?

d) Ruth is playing this game to raise $120, which she needs very badly (she owes this money to a loan shark). What strategy should she use?

e) Does Charlie have a strategy that is guaranteed to hold Ruth's winnings to less than $90?

f) Does Charlie have a strategy that is guaranteed to hold Ruth's winnings to less than $120?

g) It so happens that Charlie only has $125. Should he lose more, he will have to ask his parents again for some money, something he is not willing to do anymore. If he still wants to play, what should his strategy be?

46. The payoffs in the game below denote dollars. Ruth and Charlie are about to play this game 100 times.

3	−2
−1	2

a) Does Ruth have a strategy that guarantees her winning something each time the game is played?

b) Does Ruth have a strategy that guarantees that her losses will not exceed $120?

c) Does Ruth have a strategy that guarantees that her losses will not exceed $80?

d) Ruth is playing this game to raise $30, which she needs very badly (she owes this money to a loan shark). What strategy should she use?

e) Does Charlie have a strategy that is guaranteed to hold Ruth's winnings to less than $90?

f) Does Charlie have a strategy that is guaranteed to hold Ruth's winnings to less than $120?

g) It so happens that Charlie only has $60. Should he lose more, he will have to ask his parents again for some money, something he is not willing to do anymore. If he still wants to play, what should his strategy be?

47. The payoffs in the game below denote dollars. Ruth and Charlie are about to play this game 100 times.

3	2
−1	2

a) Does Ruth have a strategy that guarantees her winning something each time the game is played?

b) Does Ruth have a strategy that guarantees her winning a total of $200 or more?

c) Does Ruth have a strategy that guarantees her winning a total of $300 or more?

d) Ruth is playing this game to raise $30, which she needs very badly (she owes this money to a loan shark). What strategy should she use?

e) Does Charlie have a strategy that is guaranteed to hold Ruth's winnings to less than $90?

f) Does Charlie have a strategy that is guaranteed to hold Ruth's winnings to less than $250?

g) It so happens that Charlie only has $250. Should he lose more, he will have to ask his parents again for some money, something he is not willing to do anymore. If he still wants to play, what should his strategy be?

48. The payoffs in the game below denote dollars. Ruth and Charlie are about to play this game 100 times.

3	−3
−4	2

a) Does Ruth have a strategy that guarantees her winning something each time the game is played?

b) Does Ruth have a strategy that guarantees that her losses will not exceed $120?

c) Does Ruth have a strategy that guarantees that her losses will not exceed $300?

d) Charlie is playing this game to raise $30, which he needs very badly (he owes this money to a loan shark). What strategy should he use?

e) Does Charlie have a strategy that is guaranteed to hold Ruth's winnings to less than $180?

f) Does Charlie have a strategy that is guaranteed to hold Ruth's winnings to less than $250?

g)   It so happens that Ruth only has $35. Should she lose more, she will have to ask her parents again for some money, something she is not willing to do anymore. If she still wants to play, what should her strategy be?

49.   Verify the solution given for the Jamaican fishing village game. (In other words, verify that the stated strategies are indeed a Nash equilibrium of the full form of the game.)

50.   Verify that $R = C = \left[0, \dfrac{3}{5}, \dfrac{3}{5}, 0\right]$ constitute both maximin and minimax strategies for the two-finger morra game of Section 4.1. (In other words, verify that the stated strategies are indeed a Nash equilibrium of the full form of the game.)

4.6   Activities/Websites

4.6.1   Prisoner's Dilemma

Distribute to each student a form containing the following:

Name of student:

You have the option of either marking the box below or not. If both you and your opponent mark the box you both get 10 points, whereas if neither marks the box each gets 20 points. Should one player mark the box and the other not, the one who marked it gets 30 points, whereas the other gets nothing.

☐

Have the students return these forms to you, shuffle them, pair them off at random, and play each student against his/her pair-mate. Return the forms to the students *with* the payoffs after the results are recorded. Compare the students' behavior with the pure Nash equilibrium of this game. Repeat the game several times.

To make this game more meaningful, it is suggested that each player's payoff be converted to a certain amount of extra credit points.

The reason this is a variant of *prisoner's dilemma* is that the above payoffs were obtained by adding 15 to each of that game's payoffs and then doubling the sum.

4.6.2 The Job Applicants

Distribute to each student a form containing the following:

Name of student:

You have the option of either marking the box below or not. If both you and your opponent mark the box you both get 20 points, whereas if neither marks the box each gets 10 points. Should one player mark the box and the other not, the one who marked it gets 30 points, whereas the other gets 20 points.

☐

Have the students return these forms to you, shuffle them, pair them off at random, and play each student against his/her pair-mate. Return the forms to the students *with* their payoffs after the results are recorded. Compare the the students' collective behavior with the mixed Nash equilibrium of this game: $C = R = [0.2, 0.8]$. Repeat the game several times.

To make this game more meaningful, it is suggested that each player's payoff be converted to a certain amount of extra credit points.

This game is essentially the same as the *Job Applicant's* game.

4.6.3 A Zero-Sum Game (Instructor Against Class)

Before meeting your class, create your own box in your office and either mark it or not. Distribute to each student a form containing the following:

Name of student:

The student has the option of either marking the box below or not. If both the student and the instructor mark their boxes or if neither of them marks their boxes, the student gets 0 points. If the student does not and the instructor does mark, the student gets 20 points. If the student marks and the instructor does not mark, the student gets 10 points.

☐

Have the students return these forms to you and play your decision against each. Return the form to them *with* their payoff. Repeat a total of 10 times or more. Assigning to the students the role of Ruth, compare their collective behavior with the maximin strategy. Try to hold the students' gains down as much as possible.

To make this game more meaningful, it is suggested that each player's payoff be converted to a certain amount of extra credit points.

Chapter 5
Linear Programming

Outside of the physical sciences and the discipline of statistics, the most popular of mathematical applications is the topic of linear programming. This tool can be used wherever decisions are to be made that need to take into account the limited nature of all resources and the inevitability of competing requirements. It was developed by George B. Dantzig (b. 1914) during World War II to resolve the huge distribution problems faced by the armed forces in keeping the several fronts supplied with men, weapons, and supplies. The various examples and exercises of this chapter should convince the student of the wide range of applications of this tool.

5.1 Graphing Inequalities

We begin with the discussion of a problem that will motivate some definitions.

EXAMPLE 5.1.1. *Three friends joined together to form a small company that manufactures chairs and full-length sofas. The production of a chair requires 6 man-hours and the materials cost $30. The production of a sofa requires 10 man-hours and the materials cost $80. The company has 120 man-hours and a budget of $870 available each week. If the profit per chair is $45 and the profit per sofa is $80, how many of each should be produced so as to maximize the weekly profit?*

Let x denote the number of chairs and y the number of sofas to be produced. In terms of these variables the profit is

$$45x + 80y$$

The problem calls for finding the x and y that result in the largest possible profit, subject to the constraints of limited labor and limited budget. Specifically, x and y are constrained by the following inequalities:

$$\text{Labor:} \quad 6x + 10y \leq 120$$
$$\text{Budget:} \quad 30x + 80y \leq 870$$

These elements are common to all *linear programming problems*. There is always a function that needs to be maximized (or minimized, as is the case when, say, the focus is on costs). This is called the *objective function*. There are also inequalities that delimit the variables on which the objective function depends. These are the *constraints*. Mathematical problems with such objective functions and constraints are called optimization problems and are in general very difficult. Linear programs are those optimization problems whose objective function and constraints happen to be *linear* in the sense that none of the variables is multiplied by each other (or by themselves). The property of linearity makes this restricted class easier, and the consensus of the mathematical community is that the available tools solve linear programs in as efficient a manner as can be expected. A typical linear program may involve hundreds of variables and constraints, and one has been known to use as many as 800,000 variables. Not surprisingly, therefore, the solution requires both computers and efficient software packages. The fact that some of these packages sell for hundreds of thousands of dollars attests to their utility and demand.

In this text, attention is limited to linear programs with two variables only, and their solution is geometric. The inequalities are graphed, and the resulting figure is then used to determine the best choice for the values of the variables. The processes of graphing the inequalities that constitute the constraints begins with the graphing of linear equations.

The graph of an equation of the form

$$Ax + By = C$$

in the Cartesian plane is a straight line. One method for sketching this graph is to find its intersections with the axes. The x-intercept is found by substituting 0 for y and solving the resulting equation for x; the y-intercept is found by substituting 0 for x and solving the resulting equation for y.

EXAMPLE 5.1.2. *Sketch the graph of $2x + 3y = 6$*

Setting $y = 0$ results in the equation

$$2x = 6 \quad \text{or} \quad x = \frac{6}{2} = 3$$

which yields the x-intercept $(3, 0)$. Setting $x = 0$ results in the equation

$$3y = 6 \quad \text{or} \quad y = \frac{6}{3} = 2$$

which yields the y-intercept $(0, 2)$. The graph of the given equation is drawn in Figure 5.1.

There are some exceptional cases where this method fails. If $C = 0$ but neither A nor B are 0, then both the x- and y-intercepts are the origin, and so another point is needed to determine the graph. Such a point is easily obtained by substituting for x any nonzero value. Any value will do, but substituting $x = B$ avoids annoying fractions.

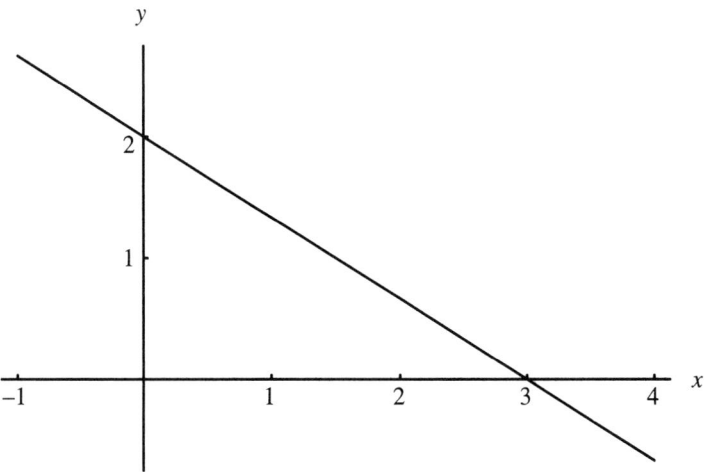

Figure 5.1   *The graph of* $2x + 3y = 6$

EXAMPLE 5.1.3. *Sketch the graph of* $2x + 3y = 0$.

Substituting either $x = 0$ or $y = 0$ yields the origin $(0, 0)$ as an intercept. The recommended substitution $x = 3$ results in the equation

$$2 \cdot 3 + 3y = 0 \quad \text{or} \quad 3y = -6 \quad \text{or} \quad y = -2$$

which yields the additional point $(3, -2)$ on the graph. The graph of the given equation is therefore the straight line joining the points $(0, 0)$ and $(3, -2)$ (Fig. 5.2).

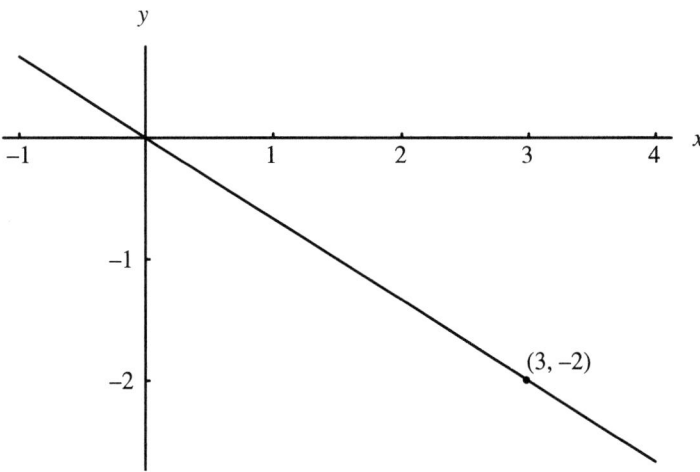

Figure 5.2   *The graph of* $2x + 3y = 0$

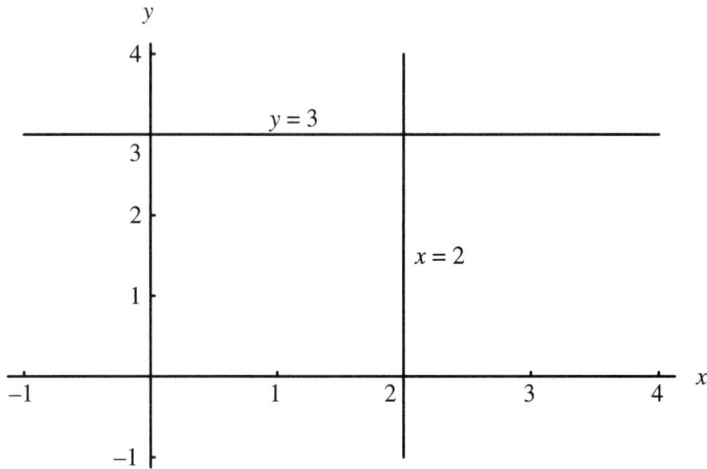

Figure 5.3   *The graphs of x = 2 and y = 3*

If $C = 0$ and exactly one of A and B is also 0, the resulting equation can be simplified to one of the two forms:

$$x = a \quad \text{or} \quad y = b$$

The first of these has the vertical line through $(a, 0)$ as its graph and the second has the horizontal line through $(0, b)$ as its graph.

EXAMPLE 5.1.4.   *Sketch the graphs of the equations* $x = 2$ *and* $y = 3$.

The graphs of these equations are displayed in Figure 5.3.

Example 5.1.1 illustrates the need for graphing inequalities. Such graphs are no longer straight lines. For example, by definition, the graph of $x \geq 2$ consists of all the points whose first coordinate is at least 2. This graph is the half-plane that lies to the right of the straight line $x = 2$ (Fig. 5.4). Similarly, because the graph of $x \leq 2$ consists

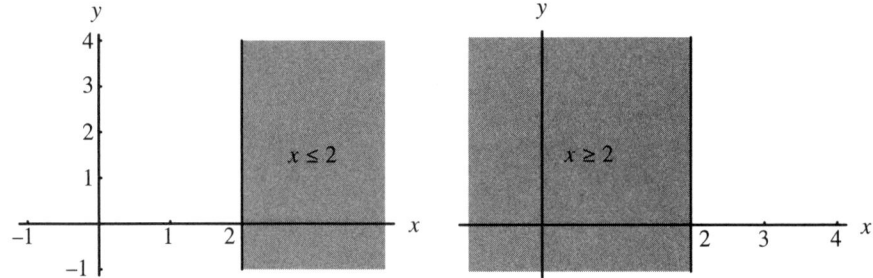

Figure 5.4   *The graphs of* $x \geq 2$ *and* $x \leq 3$

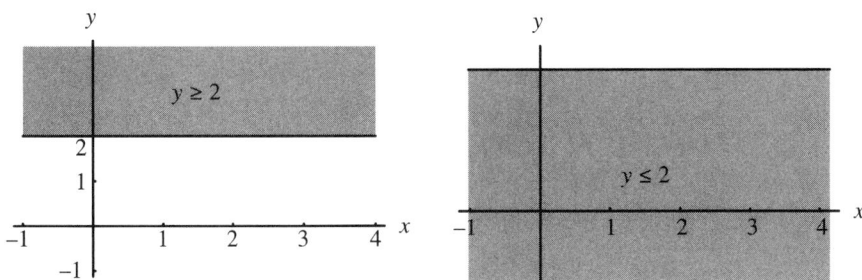

Figure 5.5 *The graphs of y ≥ 2 and y ≤ 2*

of all the points whose first coordinate is at most 2, it is the half-plane that lies to the left of the straight line $x = 2$. In general, the graphs of $x \geq a$ and $x \leq a$ are, respectively, the half-planes that lie to the right and the left of the vertical straight line $x = a$. The same reasoning, mutatis mutandis, leads to the conclusion that the graphs of $y \geq b$ and $y \leq b$ are, respectively, the half-planes that lie above and below the horizontal straight line $y = b$ (Fig. 5.5).

The graph of the inequality $x + y \geq 5$ becomes clear once it is noted (Fig. 5.6) that the sum of the coordinates of the points above and to the right of the graph of $x + y = 5$ add up to more than 5. The graph of $x + y \geq 5$ is therefore the half-plane above and to the right of the straight line of $x + y = 5$ (Fig. 5.7). Similarly, the graph of the inequality $x + y \leq 5$ is the half-plane below and to the left of the same straight line (Fig. 5.7).

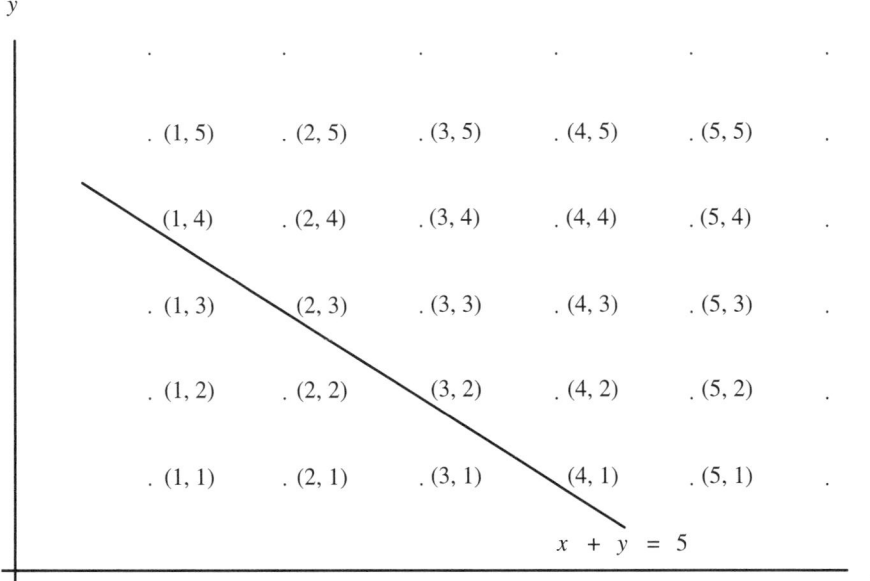

Figure 5.6 *The graph of x + y = 5*

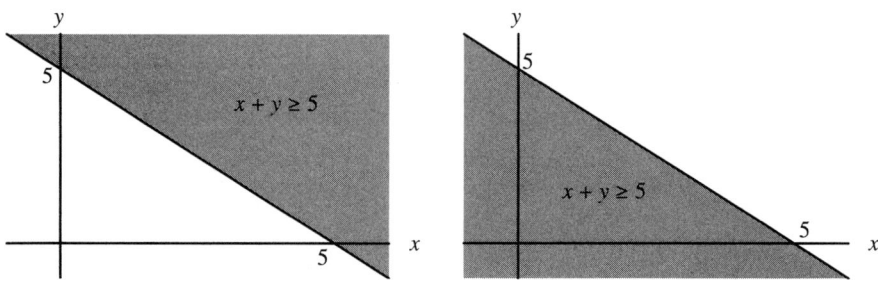

Figure 5.7   *The graphs of* $x + y \geq 5$ *and* $x + y \leq 5$

The above examples generalize to the following rule:

The graphs of the inequalities $Ax + By \geq C$ and $Ax + By \leq C$ consist of the two half-planes that lie on either side of the straight line $Ax + By = C$.

Just which of the two half-planes corresponds to which inequality depends on the signs of A, B, and C. Rather than enunciate the complicated rule, it is recommended that the reader simply pick a particular point that lies off the line and decide, by substituting its coordinates into the inequality, whether that point belongs to the graph. If the substitution results in a valid inequality, the graph consists of that half-plane that contains the point; if it does not, the graph is the other half-plane.

EXAMPLE 5.1.5.   *Sketch the graph of* $2x - 3y \geq 6$.

First, it is necessary to sketch the graph of the equation $2x - 3y = 6$. The x-intercept, obtained by setting $y = 0$, is $(3, 0)$. The y-intercept, obtained by setting $x = 0$, is $(0, -2)$. To decide which side of the line constitutes the required graph, we substitute the coordinates of the origin $(0, 0)$ into the given inequality and obtain

$$2 \cdot 0 - 3 \cdot 0 \geq 6 \quad \text{or} \quad 0 \geq 6$$

which is clearly false. This means that $(0, 0)$ is *not* in the graph of $2x - 3y \geq 6$, and hence the required graph consists of that side of the straight line that does not contain it (Fig. 5.8).

The decision to examine the origin was a matter of convenience. Had the point $(1, -1)$ been chosen instead, the substitution would have yielded

$$2 \cdot 1 - 3(-1) \geq 6 \quad \text{or} \quad 5 \geq 6$$

which is also false, leading to the same conclusion about the graph of the given inequality. Things would have turned out slightly differently, though, if the point $(3, -2)$ had been chosen. In that case substitution would have resulted in

$$2 \cdot 3 - 3(-2) \geq 6 \quad \text{or} \quad 12 \geq 6$$

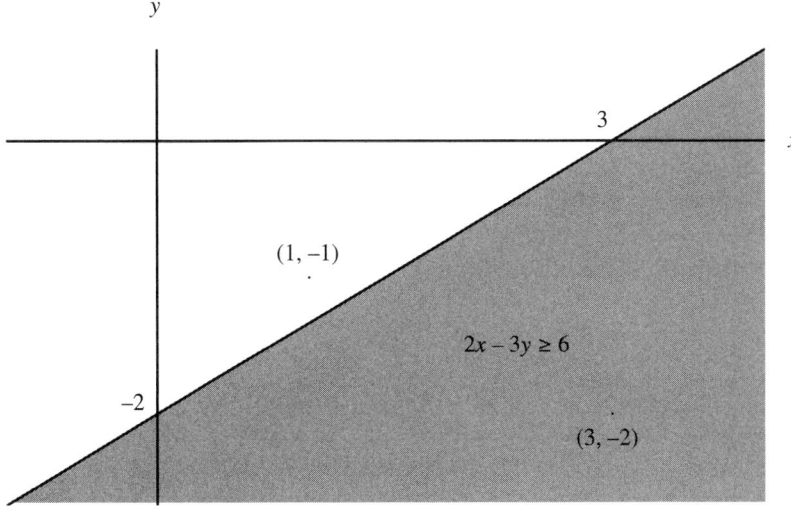

Figure 5.8 *The graph of $2x - 3y \geq 6$*

Because this is a valid inequality, this would have meant that the required half-plane must contain $(3, -2)$, which is consistent with the previous conclusion.

It is advisable to use the origin for this purpose whenever possible, that is, whenever the associated straight line does not pass through it. If the latter is the case, it is recommended that a point on either of the axes be used.

We now turn to the graphing of systems of inequalities. The graph of such a system consists of all the points that satisfy each of the inequalities.

EXAMPLE 5.1.6. *Sketch the graph of the following system:*

$$x + y \geq 6$$
$$x - 2y \leq 0$$

The straight line associated with the first inequality is $x + y = 6$, which has intercepts $(6, 0)$ and $(0, 6)$. Substitution of $x = 0$ and $y = 0$ into the inequality yields

$$0 + 0 \geq 6$$

which is false. Hence, the graph of the first inequality is the partially shaded side of $x + y = 6$ indicated in Figure 5.9.

The straight line associated with the second inequality has equation $x - 2y = 0$. Its only intercept is $(0, 0)$, but another point is obtained by substituting $y = 1$ so that

$$x - 2 \cdot 1 = 0 \quad \text{or} \quad x = 2$$

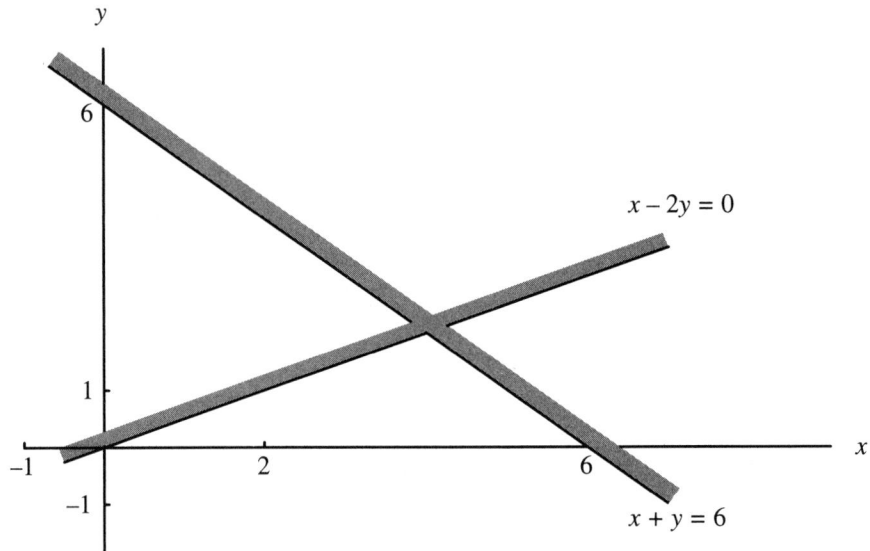

Figure 5.9   *The graphs of $x + y \geq 6$ and $x - 2y \leq 0$*

This yields the additional point $(2, 1)$ on the line. Because the point $(0, 0)$ is on this line, $(2, 0)$ is used instead to decide which of its sides constitutes the graph. The substitution of the coordinates then yields

$$2 - 2 \cdot 0 \leq 0 \quad \text{or} \quad 2 \leq 0$$

which is false. Hence, the graph of $x - 2y \leq 0$ is that side of the line $x - 2y = 0$ that does not contain the point $(2, 0)$ (Fig. 5.9).

The graphs of the two individual inequalities are sketched in Figure 5.9. The graph of the system consists of the intersection of these two graphs, namely, that one of the four regions formed by the intersection of the two straight lines both of whose sides are shaded (Fig. 5.10).

The graph of the constraints of Example 5.1.1, which was used to motivate this chapters primary concepts, is now sketched. This graph is used in Example 5.1.7 below to solve the optimization problem of Example 5.1.1.

EXAMPLE 5.1.7.   *Sketch the graph of the system of the following inequalities:*

$$6x + 10y \leq 120$$
$$30x + 80y \leq 870$$
$$x \geq 0$$
$$y \geq 0$$

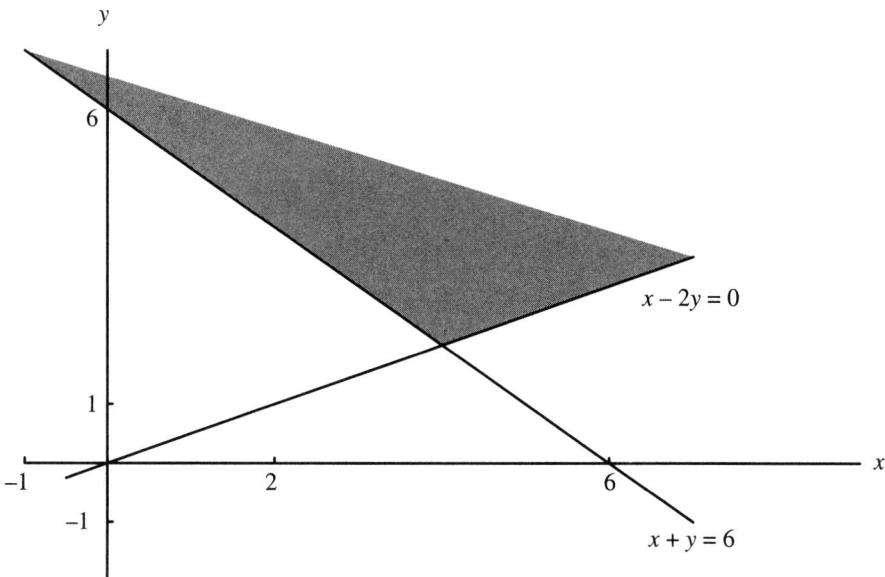

Figure 5.10   *The graph of* $x + y \geq 6$ *and* $x - Zy \leq 0$

6x + 10y ≤ 120:   The straight line $6x + 10y = 120$ has intercepts $(20, 0)$ and $(0, 12)$. The substitution of the coordinates of $(0, 0)$ into the inequality yields $6 \cdot 0 + 12 \cdot 0 \leq 120$, which is valid. Hence, the graph consists of the side that contains the origin.

30x + 80y ≤ 870:   The straight line $30x + 80y = 870$ has intercepts $(29, 0)$ and $(0, 87/8)$. The substitution of the coordinates of $(0, 0)$ into the inequality yields $30 \cdot 0 + 80 \cdot 0 \leq 870$, which is valid. Hence, the graph consists of the side of $30x + 80y = 870$, which contains the origin.

x ≥ 0:   This is the half-plane to the right of $x = 0$, which is the y-axis.

y ≥ 0:   This is the half-plane above $y = 0$, which is the x-axis.

The four individual inequalities are graphed in Figure 5.11. The graph of the system is depicted in Figure 5.12 where the vertex $(5, 9)$ was derived as follows:

$$
\begin{array}{lll}
6x + 10y = 120 & 30x + 50y = 600 & 48x + 80y = 960 \\
\quad\quad\quad\quad\quad \text{or} & 30x + 80y = 870 & 30x + 80y = 870 \\
30x + 80y = 870 & \overline{} & \overline{} \\
& -30y = -270 & 18x = 90 \\
& y = (-270)/(-30) = 9 & x = 90/18 = 5
\end{array}
$$

Figure 5.11 *Four half-planes*

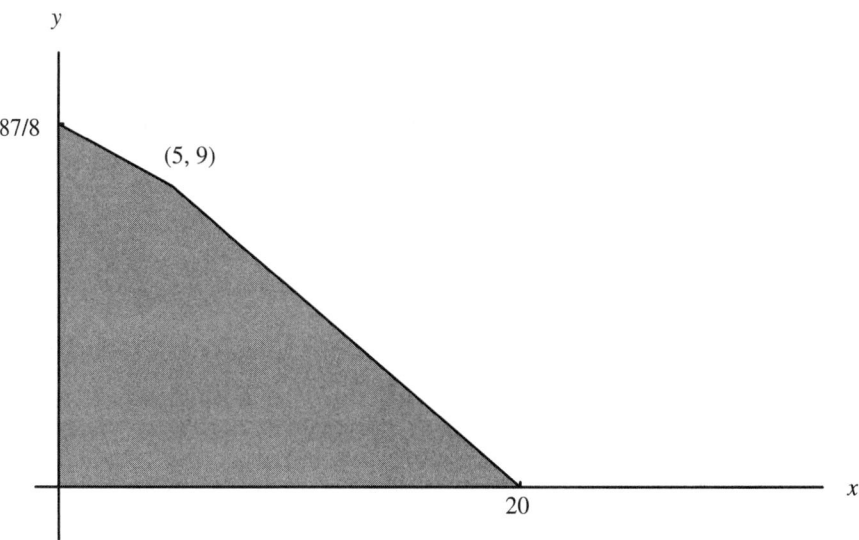

Figure 5.12 A *quadrilateral*

EXAMPLE 5.1.8. *Sketch the graph of the system of the following inequalities:*

$$x + y \leq 10$$
$$3x - 2y \geq 0$$
$$x + 6y \geq 20$$

$x + y \leq 10$: The straight line $x + y = 10$ has the intercepts $(10, 0)$ and $(0, 10)$. The substitution of the coordinates of $(0, 0)$ into the inequality yields the valid inequality $0 + 0 \leq 10$. Hence, the graph of this inequality is the side that contains the origin.

$3x - 2y \leq 0$: The straight line $3x - 2y = 0$ has the single intercept $(0, 0)$. The substitution of $x = 2$ yields the additional point $(2, 3)$. The substitution of the coordinates of $(1, 0)$ into the inequality yields the valid inequality $3 \cdot 1 - 2 \cdot 0 \geq 0$. Hence, the graph of this inequality is the side that contains the point $(1, 0)$.

$x + 6y \geq 20$: The straight line $x + 6y = 20$ has the intercepts $(0, \frac{10}{3})$ and $(20, 0)$. The substitution of the coordinates of $(0, 0)$ into the inequality yields the false inequality $0 + 6 \cdot 0 \geq 20$. Hence, the graph of this inequality is the side that does not contain the origin.

These three individual inequalities are graphed in Figure 5.13. The graph of

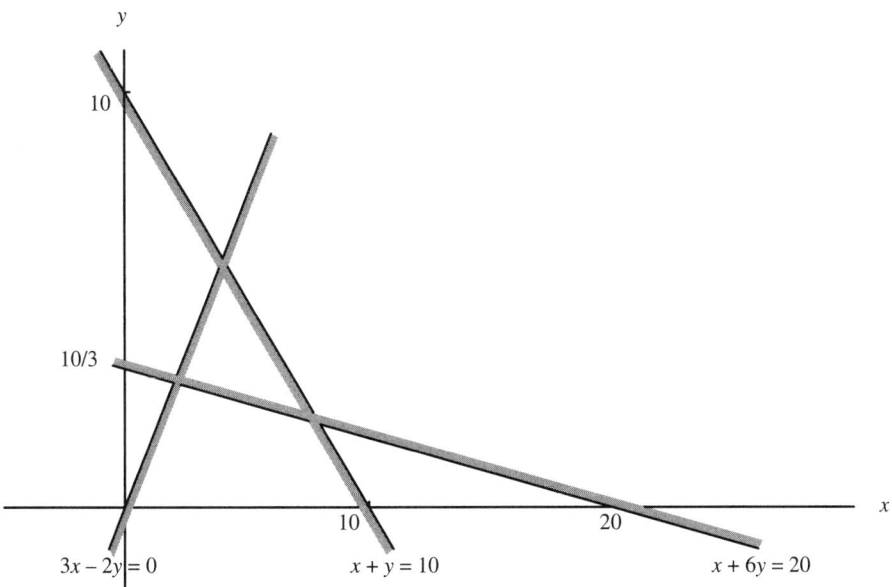

Figure 5.13 *Another four half-planes*

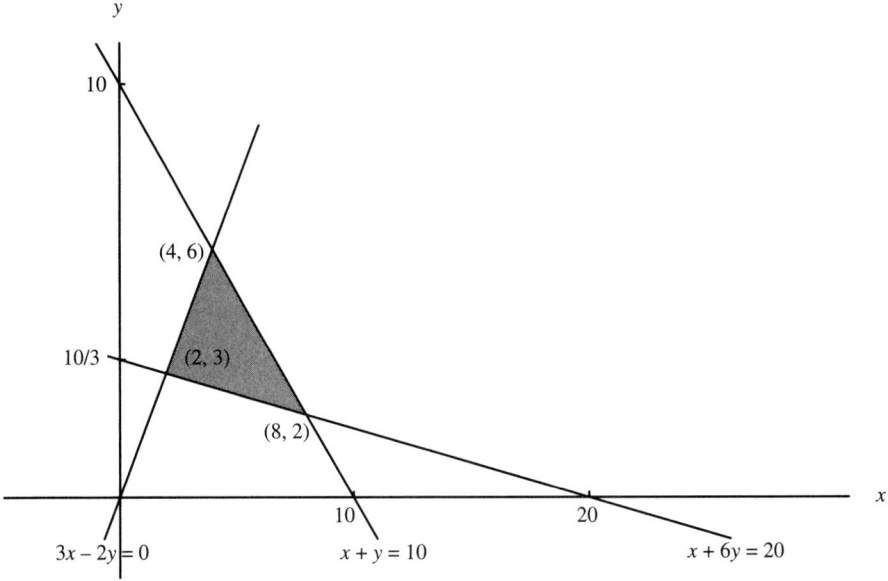

Figure 5.14 *A triangle*

the whole system consists of that region *all* of whose border lines are shaded. This is the triangle depicted in Figure 5.14. The vertices of the region also need to be computed for reasons that become clear later. The coordinates of each such vertex are found by simultaneously solving the equations of the two lines that intersect at that vertex:

(4, 6):

$$
\begin{array}{lll}
x + y = 10 & 2x + 2y = 20 & -3x - 3y = -30 \\
\quad\quad\text{or} & 3x - 2y = 0 & 3x - 2y = 0 \\
3x - 2y = 0 & \overline{\quad\quad\quad} & \overline{\quad\quad\quad} \\
& 5x = 20 & -5x = -30 \\
& x = 20/5 = 4 & x = (-30)/(-5) = 6
\end{array}
$$

(2, 3):

$$
\begin{array}{lll}
3x - 2y = 0 & 9x - 6y = 0 & 3x - 2y = 0 \\
\quad\quad\text{or} & x + 6y = 20 & -3x - 18y = -60 \\
x + 6y = 20 & \overline{\quad\quad\quad} & \overline{\quad\quad\quad} \\
& 10x = 20 & -20y = -60 \\
& x = 20/10 = 2 & x = (-60)/(-20) = 3
\end{array}
$$

(8, 2):

$$x + y = 10 \qquad -x - y = -10 \qquad -6x - 6y = -60$$
$$\text{or} \qquad\qquad \text{or}$$
$$x + 6y = 20 \qquad x + 6y = 20 \qquad x + 6y = 20$$

$$\overline{\qquad\qquad 5y = 10 \qquad\qquad -5x = -40}$$

$$y = 10/5 = 2 \qquad x = (-40)/(-5) = 8$$

EXAMPLE 5.1.9. *Sketch the graph of the system of the following inequalities:*

$$x + y \geq 10$$
$$3x - 2y \geq 0$$
$$x + 6y \geq 20$$

This example differs from the previous one only in that the first inequality has been reversed. This means that it is now necessary to use the other side of the straight line $x + y = 10$, whereas the rest of Figure 5.13 remains unchanged. We therefore obtain the preliminary sketch of Figure 5.15. The new graph is depicted in Figure 5.16.

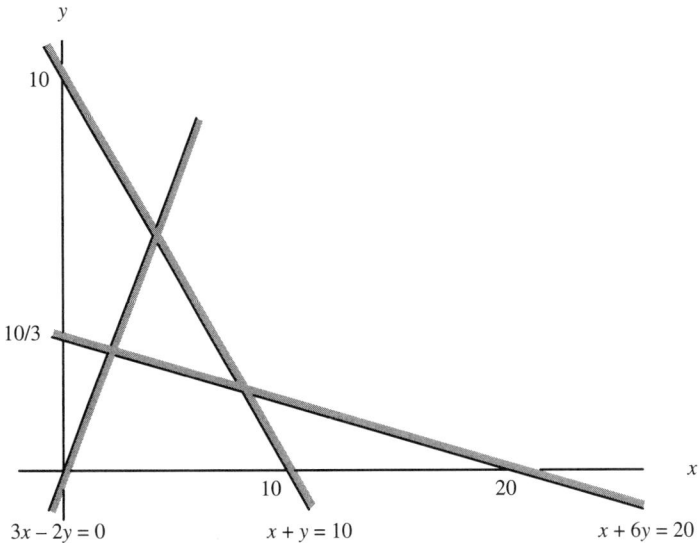

Figure 5.15 *Another four half-planes*

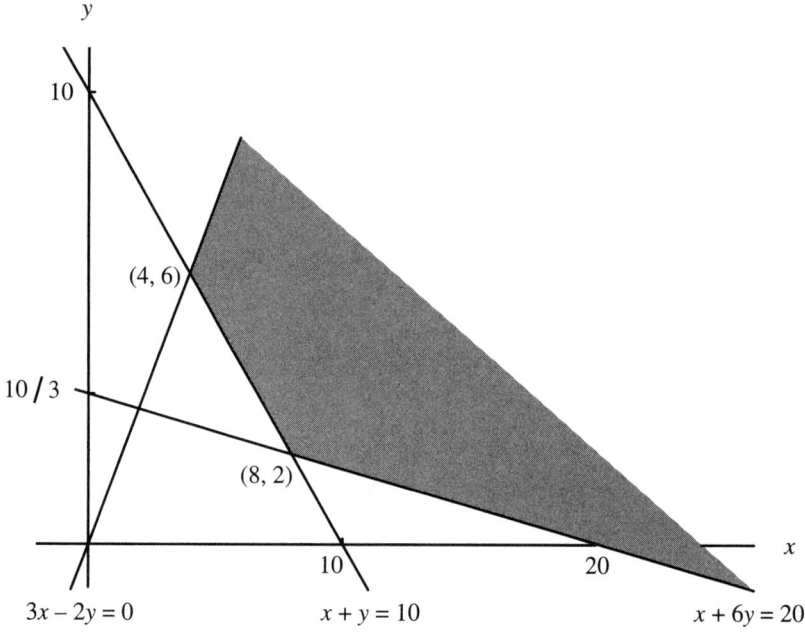

Figure 5.16 *An infinite 3-sided region*

5.1.1 Appendix: The Solution of Simultaneous Equations

Given a system of simultaneous equations, one method for obtaining its solutions is to eliminate each of the unknowns in its turn. This is accomplished by modifying the equations so that the unknown has the same coefficient in both and then subtracting the two equations. This equalization can be obtained by multiplying each equation by that unknown's coefficient in the other equation. After the subtraction, a single equation with a single unknown is left, which is easily solved.

EXAMPLE 5.1.10. *Solve the simultaneous system of the following equations:*

$$3x + 4y = 25$$
$$5x - 3y = 3$$

To eliminate the unknown x, multiply the first equation by 5, the second by 3, and subtract the resulting equations:

$$15x + 20y = 125$$
$$\underline{15x - 9y = 9}$$
$$29y = 116$$

This yields

$$y = 116/29 = 4$$

To eliminate the unknown y, multiply the first equation by -3, the second by 4, and subtract the resulting equations:

$$-9x - 12y = -75$$

$$20x - 12y = 12$$

$$\overline{-29x = -87}$$

This yields

$$x = (-87)/(-29) = 3$$

The solution to the given system, therefore, is $x = 3$ and $y = 5$.

EXERCISES 5.1

In Exercises 1–10, graph the region of solutions of the given linear inequality.

1. $3x + y \leq 6$
2. $3x + 4y \geq 12$
3. $4x - 3y \leq 12$
4. $5x - 2y \geq 10$
5. $4x - 3y \geq 0$
6. $5x - 2y \leq 0$
7. $x \geq 3$
8. $x \leq -4$
9. $y \leq -2$
10. $y \geq 2$

In Exercises 11–20, graph the region of solutions of the given system of linear inequalities. Be sure to specify the corners of the region.

11. $2x + 3y \leq 17$
    $3x - y \geq -2$
12. $5x - y \geq 6$
    $2x - 3y \leq -5$
13. $x + 2y \leq 4$
    $3x - 2y \geq -12$
    $-2x + 5y \geq 10$
14. $x - y \leq 1$
    $3x + 2y \leq -8$
    $3x - y \geq 6$
15. $x + 2y \leq 4$
    $3x - 2y \leq -12$
    $-2x + 5y \geq 10$
16. $x - y \geq 1$
    $3x + 2y \geq -8$
    $3x - y \geq 6$

17. $2x + 5y \geq 60$
    $5x + 3y \geq 60$
    $x \geq 0, y \geq 0$

18. $x + 20y \geq 460$
    $15x + y \leq 400$
    $x \geq 0, y \geq 0$

19. $x + 2y \leq 16$
    $3x + y \leq 30$
    $x + y \leq 14$
    $x \geq 0, y \geq 0$

20. $x + 3y \geq 180$
    $3x + 2y \leq 190$
    $2x + y \leq 120$
    $x \geq 0, y \geq 0$

5.2. Optimization

Having described the graph of the constraints, attention is now shifted onto the objective function. The general rule for optimizing an objective function is motivated by a simple example.

EXAMPLE 5.2.1. *Find the largest and smallest values assumed by the objective function* $x + y$ *in the quadrilateral with vertices* $(4, 2)$, $(1, 1)$, $(1, 3)$, *and* $(3, 4)$ *(Fig. 5.17).*

Figure 5.18 contains a drawing of the quadrilateral in question and also shows how the value of $x + y$ varies inside as well as in the vicinity of the quadrilateral. Note that, by definition, the value of $x + y$ along any of the lines $x + y = c$ is c and that as c assumes larger values, the graph of $x + y = c$ slides parallel to itself in the northeast direction. As long as this sliding line passes through the quadrilateral, this means that there are points (x, y) in the quadrilateral such that $x + y = c$. For those values, such as 8, for which the line $x + y = c$ misses the quadrilateral, there are no points in the quadrilateral such that $x + y = c$. It is clear from Figure 5.18 that the last value of c for which $x + y = c$ intersects the quadrilateral is $c = 7$. It follows that the largest value of $x + y$ in the quadrilateral is 7. Similarly, the first value for which the

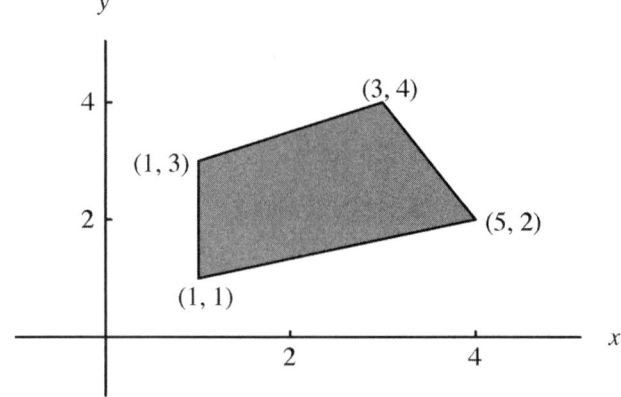

Figure 5.17   *A quadrilateral region*

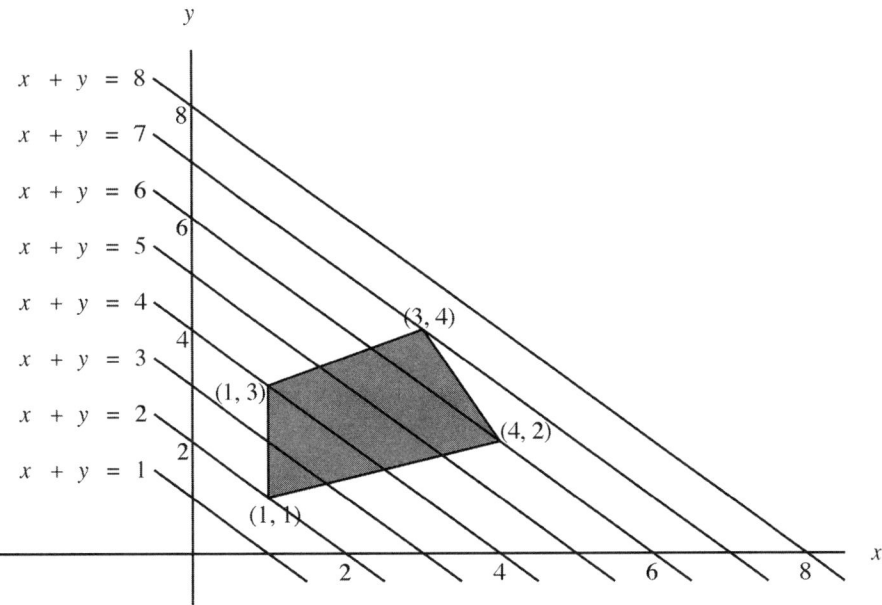

Figure 5.18 *A sweeping line*

sliding line meets the quadrilateral is $c = 2$, and hence the minimum value of $x + y$ in the quadrilateral is 2.

This particular example is typical. Varying the value of C in the general straight line

$$Ax + By = C$$

is tantamount to sliding the line parallel to itself. It is also clear that as such a straight line slides across any polygon, in its last position just as it is about to leave the polygon, it contains a vertex of that polygon. Sometimes it may contain an entire side of the polygon, but even then it still contains one of the vertices. Consequently, we have the following *vertex principle*:

> The maximum and minimum values of an objective function of the form $Ax + By$ in a polygon are assumed at some of that polygon's vertices.

This principle reduces the problem of finding the optimal value of an objective function over an entire polygon to the apparently simpler task of computing its values at the polygon's vertices. The phrase "apparently simpler" is used because in actual problems, where the number of equations and variables reaches into the hundreds, the number of vertices is huge and may easily exceed the number of atoms in the universe. Checking the value of the objective function at all these vertices is impractical, even with computers, and the main task of any linear programming software package is to find an efficient way of locating the optimizing vertices. Of course, this text's examples and exercises contain only a handful of vertices.

EXAMPLE 5.2.2. *Find the maximum and minimum values of the objective function* $2x - 3y$ *over the pentagon with vertices* $(1, 3)$, $(3, 4)$, $(5, 3)$, $(3, 1)$, *and* $(2, 1)$.

To find the answer we tabulate the values of $2x - 3y$ at the given vertices:

Vertex	$2x - 3y$
$(1, 3)$	$2 \cdot 1 - 3 \cdot 3 = 2 - 9 = -7$
$(3, 4)$	$2 \cdot 3 - 3 \cdot 4 = 6 - 12 = -6$
$(5, 3)$	$2 \cdot 5 - 3 \cdot 3 = 10 - 9 = 1$
$(3, 1)$	$2 \cdot 3 - 3 \cdot 1 = 6 - 3 = 3$
$(2, 1)$	$2 \cdot 2 - 3 \cdot 1 = 4 - 3 = 1$

Consequently, the maximum and minimum values of the objective function are 3 and −7, respectively. Figure 5.19 displays both the polygon and the behavior of the objective function relative to this polygon. Of course, the vertex principle obviates the need for such diagrams.

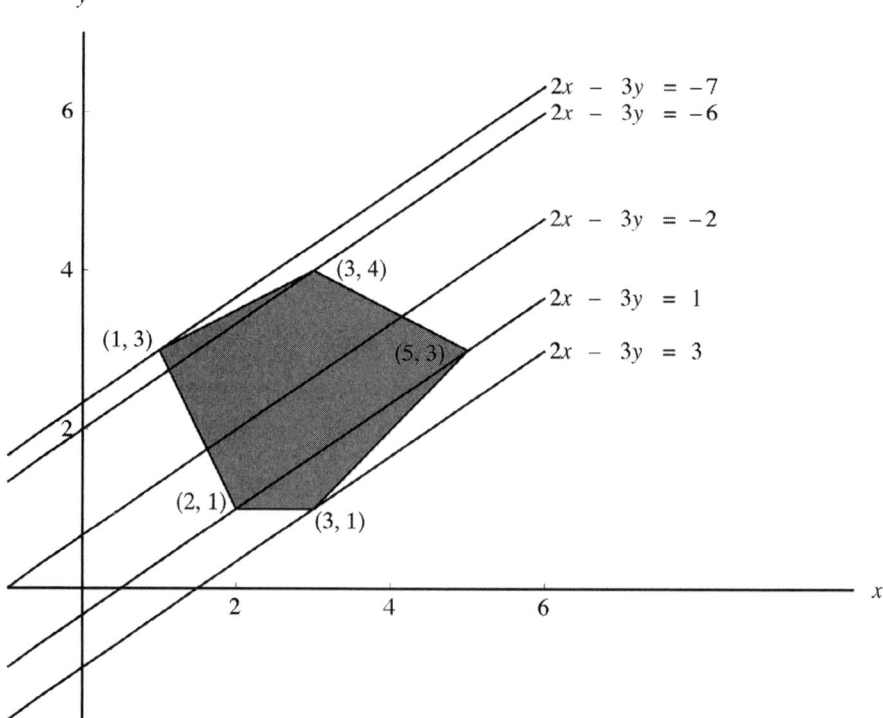

Figure 5.19   *Another sweeping line*

EXAMPLE 5.2.3. *Find the maximum and minimum values of the objective function* $x + 2y$ *over the pentagon with vertices* $(1, 3), (3, 4), (5, 3), (3, 1)$, *and* $(2, 1)$.

To find the answer we tabulate the values of $x + 2y$ at the given vertices:

Vertex	$x + 2y$
$(1, 3)$	$1 + 2 \cdot 3 = 1 + 6 = 7$
$(3, 4)$	$3 + 2 \cdot 4 = 3 + 8 = 11$
$(5, 3)$	$5 + 2 \cdot 3 = 5 + 6 = 11$
$(3, 1)$	$3 + 2 \cdot 1 = 3 + 2 = 5$
$(2, 1)$	$2 + 2 \cdot 1 = 2 + 2 = 4$

Consequently, the maximum and minimum values of the objective function are 11 and 4, respectively. Note that the maximum occurs at two points. Figure 5.20 displays both the polygon and the behavior of the objective function relative to this polygon. This diagram makes it clear that the maximum value of $x + 2y$ is actually assumed at every point on the side of the polygon that joins the vertices $(3, 4)$ and $(5, 3)$.

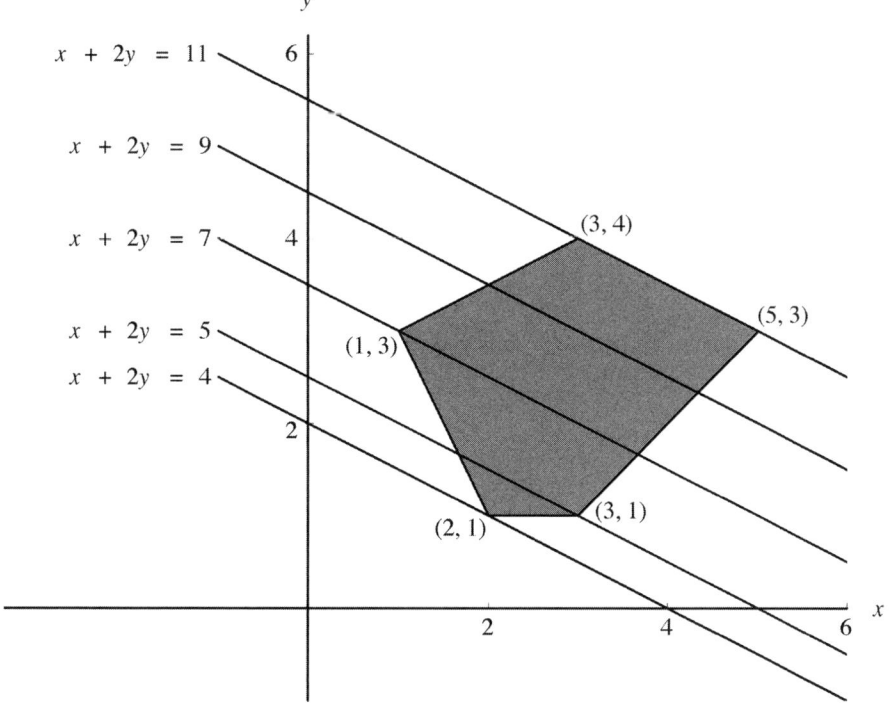

Figure 5.20   *More than one maximizing point*

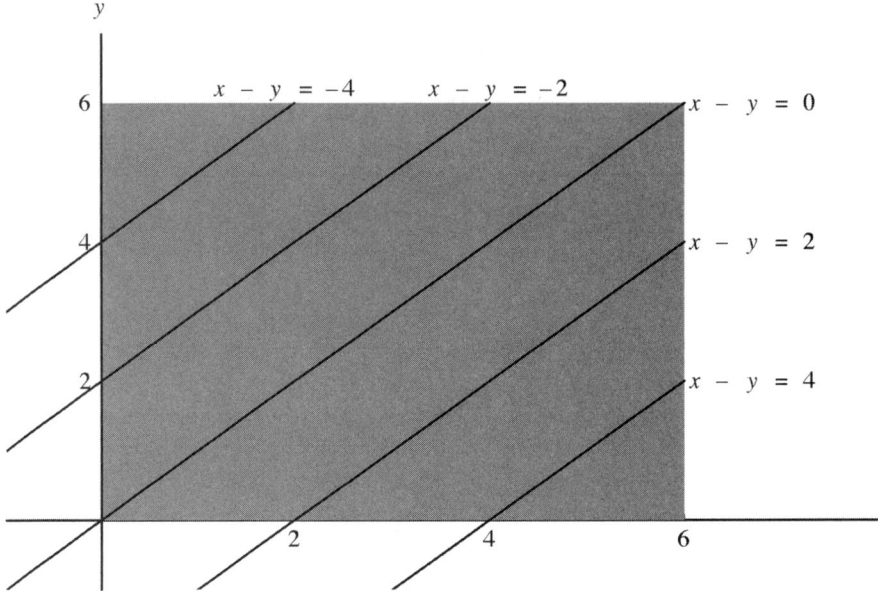

Figure 5.21   *An objective function with no maximum and no minimum*

It was seen in Example 5.1.9 that the graphs of systems of inequalities can have infinite extents. In such cases, objective functions may not have either maxima or minima. For instance, Figure 5.21 demonstrates that the objective function $x - y$ has neither a maximum nor a minimum over the graph of the system $x \geq 0$, $y \geq 0$, which is the first quadrant. The reason for this is that for every value of c, the straight line $x - y = c$, which has a southwest to northeast direction, is bound to intersect the first quadrant at some point.

The precise explanation of which objective functions have maxima or minima over infinitely extended regions falls beyond the scope of this text. For the purpose of the subsequent applications, the following rule suffices:

> If an infinitely extended polygon lies completely in the first quadrant and A and B are positive numbers, then the objective function $Ax + By$ has a minimum and this minimum is assumed at one of the polygon's vertices.

EXAMPLE 5.2.4. *Find the minimum value of $2x + 5y$ over the region depicted in Figure 5.22.*

According to the above rule such a minimum value exists, and it is assumed at one of the specified vertices. Straightforward substitution yields the table

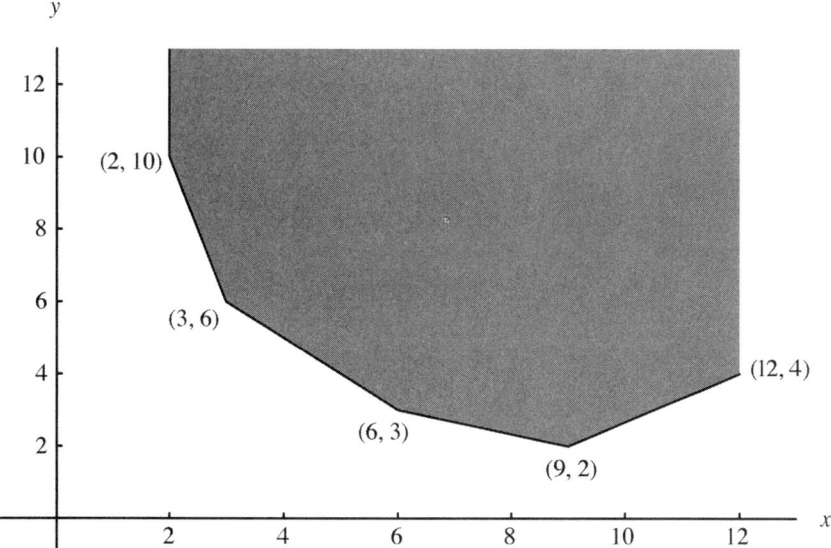

Figure 5.22 An infinite 5-sided region

Vertex	$2x + 5y$
$(2, 10)$	$2 \cdot 2 + 5 \cdot 10 = 4 + 50 = 54$
$(3, 6)$	$2 \cdot 3 + 5 \cdot 6 = 6 + 30 = 36$
$(6, 3)$	$2 \cdot 6 + 5 \cdot 3 = 12 + 15 = 27$
$(9, 2)$	$2 \cdot 9 + 5 \cdot 2 = 18 + 10 = 28$
$(12, 4)$	$2 \cdot 12 + 5 \cdot 4 = 24 + 20 = 44$

Hence, the minimum value of this objective function is 27.

We now turn to the solution of some fairly realistic word problems.

EXAMPLE 5.2.5. *Recall the small furniture factory of Example 5.1.1 whose constraints were graphed in Example 5.1.7 and whose objective function is 45x + 80y. These lead to the following table:*

Vertex	$45x + 80y$
$(0, 0)$	$45 \cdot 0 + 80 \cdot 0 = 0$
$(20, 0)$	$45 \cdot 20 + 80 \cdot 0 = 900$
$(5, 9)$	$45 \cdot 5 + 80 \cdot 9 = 945$
$\left(0, \dfrac{87}{8} \right)$	$45 \cdot 0 + 80 \cdot \dfrac{87}{8} = 870$

Because 945 is the largest of the computed values of the objective function, it follows that the company will optimize its profits by manufacturing five chairs and nine sofas weekly.

EXAMPLE 5.2.6. *A veterinarian wishes to prepare a new cat food that consists of a mixture of two ingredients, I and II. Each unit of ingredient I costs $3 and contains 2 grams of protein and 4 grams of carbohydrates. Each unit of ingredient II costs $4 and contains 4 grams of protein and 5 grams of carbohydrates. Determine the least expensive mixture the veterinarian can prepare, subject to the constraints that it must contain at least 26 grams of protein and 40 grams of carbohydrates.*

The objective function is the cost of the mixture. Because this cost depends on the amounts of foods I and II that are used in the mixture, these amounts are denoted by x and y, respectively, and in terms of these variables the objective function is

$$3x + 4y$$

To obtain the constraints, it is helpful to organize the information in the problem in terms of a tableau that lists on the left those elements of the problem on which the objective function depends directly. Each of these in turn depends on some other resources that should be listed at the top of the tableau. The actual dependencies become the entries of the array. The tableau's bottom row should list the various constraints on the resources.

	Protein	Carbohydrates
Ingredient I (x)	2	4
Ingredient II (y)	4	5
Constraint	≥ 26	≥ 40

The constraint inequalities are then easily read off the tableau:

$$2x + 4y \geq 26$$
$$4x + 5y \geq 40$$
$$x \geq 0, y \geq 0$$

This yields the preliminary diagram of Figure 5.23. To find the missing vertex we solve simultaneously

$$
\begin{array}{ccccc}
2x + 4y = 26 & & 4x + 8y = 52 & & 10x + 20y = 130 \\
\text{or} & & \text{or} & & \\
4x + 5y = 40 & & 4x + 5y = 40 & & 16x + 20y = 160 \\
\hline
 & & 3y = 12 & & -6x = -30 \\
 & & y = 12/3 = 4 & & x = (-30)/(-6) = 5
\end{array}
$$

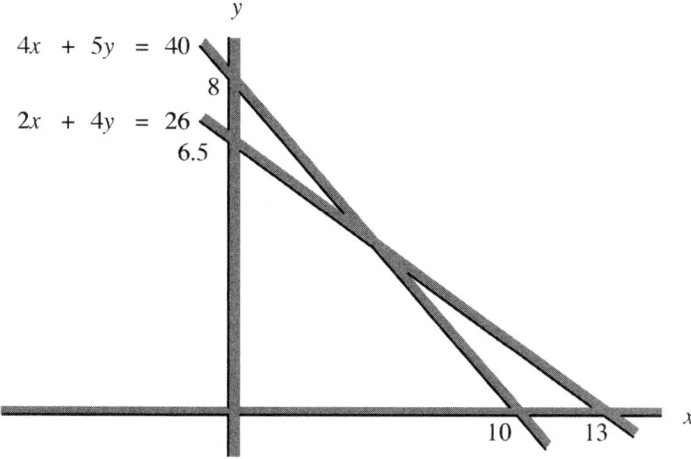

Figure 5.23 *Four half-planes*

and obtain the vertex coordinates $(5, 4)$. The feasible region is depicted in Figure 5.24. The evaluation of the objective function at the three vertices of the region yields the following table:

Vertex	$3x + 4y$
$(0, 8)$	$3 \cdot 0 + 4 \cdot 8 = 32$
$(5, 4)$	$3 \cdot 5 + 4 \cdot 4 = 31$
$(13, 0)$	$3 \cdot 13 + 4 \cdot 0 = 39$

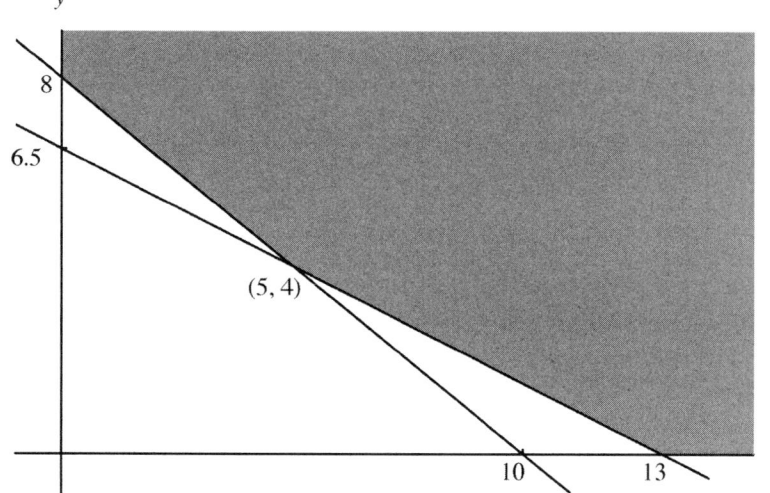

Figure 5.24 *An infinite 4-sided region*

Hence, the minimal cost of the dietitian's mixture is $31, and it is obtained by mixing 5 units of food I with 4 units of food II.

EXAMPLE 5.2.7. *Suppose Example 5.2.6 is modified by stipulating that the cost of each unit of food I is $5 rather than $3. How does this affect the problem's solution?*

Because the constraints are unchanged, Figure 5.24 still depicts the feasible region. On the other hand, the objective function is now $3x + 5y$ that, when evaluated at the vertices, yields the following table:

Vertex	$5x + 4y$
$(0, 8)$	$5 \cdot 0 + 4 \cdot 8 = 32$
$(5, 4)$	$5 \cdot 5 + 4 \cdot 4 = 41$
$(13, 0)$	$5 \cdot 13 + 4 \cdot 0 = 65$

The minimum cost is therefore $32, which is realized by using food II exclusively. Evidently, food I has become too expensive.

EXERCISES 5.2

1. Find the maximum and minimum values of $x + 10y$ over the quadrilateral with vertices $(2, 0)$, $(0, 2)$, $(1, 10)$, $(9, 2)$.

2. Find the maximum and minimum values of $2x - 3y$ over the quadrilateral with vertices $(2, 0)$, $(0, 2)$, $(1, 10)$, $(9, 2)$.

3. Find the maximum and minimum values of $10x + y$ over the quadrilateral with vertices $(2, 0)$, $(0, 2)$, $(1, 10)$, $(9, 2)$.

4. Find the maximum and minimum values of $3x - 2y$ over the quadrilateral with vertices $(2, 0)$, $(0, 2)$, $(1, 10)$, $(9, 2)$.

5. The Johnson nut company can obtain from its suppliers 1000 pounds of almonds and 2000 pounds of cashews that produces two mixtures, X and Y. Mix X uses 2 parts almonds to 1 part of cashews. Mix Y uses 1 part almonds to 3 parts cashews. How much of each mix should the Johnson company produce if it wishes to maximize its profit and each pound of the mixes brings in the following profit:

   a) X $5, Y $3

   b) X $4, Y $3

   c) X $1, Y $4

6. Redo the previous exercise under the additional assumption that at least 400 pounds of each mix must be produced.

7. A small furniture company manufactures two products, tables and chairs. The production of one table calls for 80 minutes of labor and $36 worth of materials. On the other hand, each chair requires only 40 minutes of labor and $24 worth of materials. Each employee is, of course, to work only 40 hours each week. Budgetary constraints limit the spending of each employee to $1,260. Determine the maximum weekly profit if the individual items yield the following profits:

   a) Tables $40, chairs $35

   b) Tables $60, chairs $35

   c) Tables $100, chairs $35

8. The Bay store sells two blends of coffee, the House blend and the Shock You blend. The first blend consists of 2 parts Colombian beans to 1 part Hawaiian beans. The second blend uses 1 part Colombian beans to 3 parts Hawaiian beans. The daily supply of each of the two types of beans is limited to 50 pounds each. How many pounds of each blend should be mixed and sold so as to maximize the daily profit if the profits per pound of the blends are:

   a) House $2, Shock You $2

   b) House $3, Shock You $1

   c) House $2, Shock You $1

9. The How-Now-Brown-Cow company produces 10,000 containers of cottage cheese per week that it sells to two chains of supermarkets, Zigzag and Syzygy. The former supermarket needs at least 4000 containers each week, whereas the latter supermarket needs at least 3000 containers each week. Help the How-Now-Brown-Cow company minimize its total shipping costs if the average shipping cost per container is:

   a) 5 cents to Zigzag, 4 cents to Syzygy

   b) 5 cents to Zigzag, 6 cents to Syzygy

   c) 5 cents to Zigzag, 7 cents to Syzygy

10. The Crying-Cow company produces 11,000 containers of cottage cheese per week that it sells to two chains of supermarkets, Zigzag and Syzygy. Each of the chains requires between 4000 and 6000 containers per week. Help the How-Now-Brown-Cow company schedule its shipments if the average shipping cost per container is:

   a) 5 cents to Zigzag, 4 cents to Syzygy

   b) 5 cents to Zigzag, 6 cents to Syzygy

   c) 5 cents to Zigzag, 7 cents to Syzygy

11. Gail Sturm has $120,000 to invest in two types of securities, A (high risk) and B (low risk). The broker gives Gail the following advice that he feels has served his customers well in the past; invest at least one fourth of your funds in type A securities and, at most, one third in type B. Since Gail is a risk taker, she decides to invest more in the type A securities than in type B securities. How should Gail distribute her investments if the securities are expected to have the following annual yields:

   a)   A 15%, B 6%

   b)   A 12%, B 8%

   c)   A 12%, B 15%

12. Joe has a weekly advertising budget of $6,000 that he can use for either local TV or local radio advertising. A radio ad costs $200 whereas a TV ad costs $500. As Joe does not wish to antagonize anyone, he decides to invest at least $1,000 in each advertising venue. How should Joe distribute his funds so as to reach the widest possible audience if ads on radio and TV reach the following numbers of people, respectively:

   a)   Radio 1000, TV 20,000

   b)   Radio 1000, TV 5000

   c)   Radio 1000, TV 2000

13. The Turing company produces printers and scanners. The cost of materials is $10 for each printer and $20 for each scanner. The labor for each printer costs $16 and the labor for each scanner costs $8. The weekly materiel budget and labor budgets are $12,000 each. Determine a production schedule that will maximize the profits if the individual items yield the following profits:

   a)   Printer $10, scanner $25

   b)   Printer $25, scanner $20

   c)   Printer $12, scanner $20

14. The Chew dog food company produces a healthy brand of pet food that is a mixture of two foods, say Tori and Lou-Anne. Each unit of the Tori food contains 4 ounces of protein and 6 ounces of carbohydrates. Each unit of the Lou-Anne food contains 6 ounces of protein and 3 ounces of carbohydrates. If the company standards require that the mixture contain at least 42 ounces of protein and 45 ounces of carbohydrates, what is the least expensive mixture if the unit price of the foods are:

   a)   Tori $4, Lou-Anne $2

   b)   Tori $3, Lou-Anne $3

   c)   Tori $2, Lou-Anne $4

15. World Motors manufactures quarter-ton, half-ton, and three-quarter-ton panel trucks. Express Delivery Service has placed an order for at least 300 quarter-ton, 450 half-ton, and 450 three-quarter-ton panel trucks. World Motors builds the trucks at two plants, one in Detroit and one in Los Angeles. The Detroit plant produces 30 quarter-ton trucks, 60 half-ton trucks, and 90 three-quarter-ton trucks each week at a total cost of $540,000. The Los Angeles plant produces 60 quarter-ton trucks, 45 half-ton trucks, and 30 three-quarter-ton trucks each week at a total cost of $360,000. How should World Motors schedule its two plants so that it can fill this order at minimum cost? What is that minimum cost? How would the answer change if each plant cost $500,000 per week?

16. Barton's Chocolates produces semisweet chocolate chips and milk chocolate chips at its plants in Bay City and Eustacia. The Bay City plant produces 3000 pounds of semisweet chips and 2000 pounds of milk chocolate chips each day at a cost of $1100, whereas the Eustacia plant produces 1000 pounds of semisweet chips and 6000 pounds of milk chocolate chips each day at a cost of $1800. Barton's has an order from SuperConvenience Stores for at least 30,000 pounds of semisweet chips and at least 60,000 pounds of milk chocolate chips. How should it schedule its production so that it can fulfill the order at minimum cost? What is that minimum cost?

Appendix

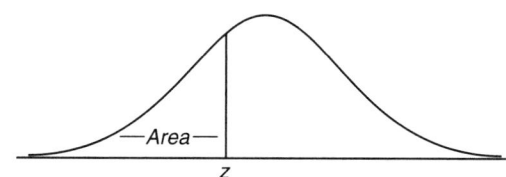

Areas under the normal curve

z	0.00	0.01	0.02	0.03	0.04	0.05	0.06	0.07	0.08	0.09
−3.0	0.0013	0.0014	0.0014	0.0015	0.0015	0.0016	0.0016	0.0017	0.0018	0.0018
−2.9	0.0019	0.0019	0.0020	0.0021	0.0021	0.0022	0.0023	0.0023	0.0024	0.0025
−2.8	0.0026	0.0026	0.0027	0.0028	0.0029	0.0030	0.0031	0.0032	0.0033	0.0034
−2.7	0.0035	0.0036	0.0037	0.0038	0.0039	0.0040	0.0041	0.0043	0.0044	0.0045
−2.6	0.0047	0.0048	0.0049	0.0051	0.0052	0.0054	0.0055	0.0057	0.0059	0.0060
−2.5	0.0062	0.0064	0.0066	0.0068	0.0069	0.0071	0.0073	0.0075	0.0078	0.0080
−2.4	0.0082	0.0084	0.0087	0.0089	0.0091	0.0094	0.0096	0.0099	0.0102	0.0104
−2.3	0.0107	0.0110	0.0113	0.0116	0.0119	0.0122	0.0125	0.0129	0.0132	0.0136
−2.2	0.0139	0.0143	0.0146	0.0150	0.0154	0.0158	0.0162	0.0166	0.0170	0.0174
−2.1	0.0179	0.0183	0.0188	0.0192	0.0197	0.0202	0.0207	0.0212	0.0217	0.0222
−2.0	0.0228	0.0233	0.0239	0.0244	0.0250	0.0256	0.0262	0.0268	0.0274	0.0281
−1.9	0.0287	0.0294	0.0301	0.0307	0.0314	0.0322	0.0329	0.0336	0.0344	0.0351
−1.8	0.0359	0.0367	0.0375	0.0384	0.0392	0.0401	0.0409	0.0418	0.0427	0.0436
−1.7	0.0446	0.0455	0.0465	0.0475	0.0485	0.0495	0.0505	0.0516	0.0526	0.0537
−1.6	0.0548	0.0559	0.0571	0.0582	0.0594	0.0606	0.0618	0.0630	0.0643	0.0655
−1.5	0.0668	0.0681	0.0694	0.0708	0.0721	0.0735	0.0749	0.0764	0.0778	0.0793
−1.4	0.0808	0.0823	0.0838	0.0853	0.0869	0.0885	0.0901	0.0918	0.0934	0.0951
−1.3	0.0968	0.0985	0.1003	0.1020	0.1038	0.1056	0.1075	0.1093	0.1112	0.1131
−1.2	0.1151	0.1170	0.1190	0.1210	0.1230	0.1251	0.1271	0.1292	0.1314	0.1335
−1.1	0.1357	0.1379	0.1401	0.1423	0.1446	0.1469	0.1492	0.1515	0.1539	0.1562
−1.0	0.1587	0.1611	0.1635	0.1660	0.1685	0.1711	0.1736	0.1762	0.1788	0.1814
−0.9	0.1841	0.1867	0.1894	0.1922	0.1949	0.1977	0.2005	0.2033	0.2061	0.2090

Table A. Cumulative normal probabilities (*continues*)

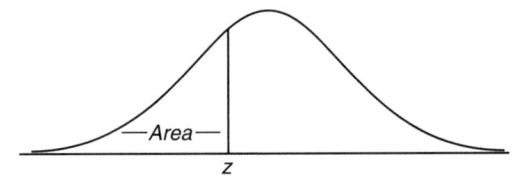

Areas under the normal curve

z	0.00	0.01	0.02	0.03	0.04	0.05	0.06	0.07	0.08	0.09
−0.8	0.2119	0.2148	0.2177	0.2206	0.2236	0.2266	0.2296	0.2327	0.2358	0.2389
−0.7	0.2420	0.2451	0.2483	0.2514	0.2546	0.2578	0.2611	0.2643	0.2676	0.2709
−0.6	0.2743	0.2776	0.2810	0.2843	0.2877	0.2912	0.2946	0.2981	0.3015	0.3050
−0.5	0.3085	0.3121	0.3156	0.3192	0.3228	0.3264	0.3300	0.3336	0.3372	0.3409
−0.4	0.3446	0.3483	0.3520	0.3557	0.3594	0.3632	0.3669	0.3707	0.3745	0.3783
−0.3	0.3821	0.3859	0.3897	0.3936	0.3974	0.4013	0.4052	0.4090	0.4129	0.4168
−0.2	0.4207	0.4247	0.4286	0.4325	0.4364	0.4404	0.4443	0.4483	0.4522	0.4562
−0.1	0.4602	0.4641	0.4681	0.4721	0.4761	0.4801	0.4840	0.4880	0.4920	0.4960
0.0	0.5000	0.5040	0.5080	0.5120	0.5160	0.5199	0.5239	0.5279	0.5319	0.5359
0.1	0.5398	0.5438	0.5478	0.5517	0.5557	0.5596	0.5636	0.5675	0.5714	0.5753
0.2	0.5793	0.5832	0.5871	0.5910	0.5948	0.5987	0.6026	0.6064	0.6103	0.6141
0.3	0.6179	0.6217	0.6255	0.6293	0.6331	0.6368	0.6406	0.6443	0.6480	0.6517
0.4	0.6554	0.6591	0.6628	0.6664	0.6700	0.6736	0.6772	0.6808	0.6844	0.6879
0.5	0.6915	0.6950	0.6985	0.7019	0.7054	0.7088	0.7123	0.7157	0.7190	0.7224
0.6	0.7257	0.7291	0.7324	0.7357	0.7389	0.7422	0.7454	0.7486	0.7517	0.7549
0.7	0.7580	0.7611	0.7642	0.7673	0.7704	0.7734	0.7764	0.7794	0.7823	0.7852
0.8	0.7881	0.7910	0.7939	0.7967	0.7995	0.8023	0.8051	0.8078	0.8106	0.8133
0.9	0.8159	0.8186	0.8212	0.8238	0.8264	0.8289	0.8315	0.8340	0.8365	0.8389
1.0	0.8413	0.8438	0.8461	0.8485	0.8508	0.8531	0.8554	0.8577	0.8599	0.8621
1.1	0.8643	0.8665	0.8686	0.8708	0.8729	0.8749	0.8770	0.8790	0.8810	0.8830
1.2	0.8849	0.8869	0.8888	0.8907	0.8925	0.8944	0.8962	0.8980	0.8997	0.9015
1.3	0.9032	0.9049	0.9066	0.9082	0.9099	0.9115	0.9131	0.9147	0.9162	0.9177
1.4	0.9192	0.9207	0.9222	0.9236	0.9251	0.9265	0.9279	0.9292	0.9306	0.9319
1.5	0.9332	0.9345	0.9357	0.9370	0.9382	0.9394	0.9406	0.9418	0.9429	0.9441
1.6	0.9452	0.9463	0.9474	0.9484	0.9495	0.9505	0.9515	0.9525	0.9535	0.9545
1.7	0.9554	0.9564	0.9573	0.9582	0.9591	0.9599	0.9608	0.9616	0.9625	0.9633
1.8	0.9641	0.9649	0.9656	0.9664	0.9671	0.9678	0.9686	0.9693	0.9699	0.9706
1.9	0.9713	0.9719	0.9726	0.9732	0.9738	0.9744	0.9750	0.9756	0.9761	0.9767
2.0	0.9772	0.9778	0.9783	0.9788	0.9793	0.9798	0.9803	0.9808	0.9812	0.9817
2.1	0.9821	0.9826	0.9830	0.9834	0.9838	0.9842	0.9846	0.9850	0.9854	0.9857
2.2	0.9861	0.9864	0.9868	0.9871	0.9875	0.9878	0.9881	0.9884	0.9887	0.9890
2.3	0.9893	0.9896	0.9898	0.9901	0.9904	0.9906	0.9909	0.9911	0.9913	0.9916

Table A.   Cumulative normal probabilities (*continues*)

Areas under the normal curve

z	0.00	0.01	0.02	0.03	0.04	0.05	0.06	0.07	0.08	0.09
2.4	0.9918	0.9920	0.9922	0.9925	0.9927	0.9929	0.9931	0.9932	0.9934	0.9936
2.5	0.9938	0.9940	0.9941	0.9943	0.9945	0.9946	0.9948	0.9949	0.9951	0.9952
2.6	0.9953	0.9955	0.9956	0.9957	0.9959	0.9960	0.9961	0.9962	0.9963	0.9964
2.7	0.9965	0.9966	0.9967	0.9968	0.9969	0.9970	0.9971	0.9972	0.9973	0.9974
2.8	0.9974	0.9975	0.9976	0.9977	0.9977	0.9978	0.9979	0.9979	0.9980	0.9981
2.9	0.9981	0.9982	0.9982	0.9983	0.9984	0.9984	0.9985	0.9985	0.9986	0.9986
3.0	0.9987	0.9987	0.9987	0.9988	0.9988	0.9989	0.9989	0.9989	0.9990	0.9990

Table A. Cumulative normal probabilities

n = 2

					p				
k	0.1	0.2	03	0.4	0.5	0.6	0.7	0.8	0.9
0	0.8100	0.6400	0.4900	0.3600	0.2500	0.1600	0.0900	0.0400	0.0100
1	0.9900	0.9600	0.9100	0.8400	0.7500	0.6400	0.5100	0.3600	0.1900
02	1.0000	1.0000	1.0000	1.0000	1.0000	1.0000	1.0000	1.0000	1.0000

n = 3

					p				
k	0.1	0.2	03	0.4	0.5	0.6	0.7	0.8	0.9
0	0.7290	0.5120	0.3430	0.2160	0.1250	0.0640	0.0270	0.0080	0.0010
1	0.9720	0.8960	0.7840	0.6480	0.5000	0.3520	0.2160	0.1040	0.0280
2	0.9990	0.9920	0.9730	0.9360	0.8750	0.7840	0.6570	0.4880	0.2710
3	1.0000	1.0000	1.0000	1.0000	1.0000	1.0000	1.0000	1.0000	1.0000

n = 4

					p				
k	0.1	0.2	03	0.4	0.5	0.6	0.7	0.8	0.9
0	0.6561	0.4096	0.2401	0.1296	0.0625	0.0256	0.0081	0.0016	0.0000
1	0.9477	0.8192	0.6517	0.4752	0.3125	0.1792	0.0837	0.0272	0.0037
2	0.9963	0.9728	0.9163	0.8208	0.6875	0.5248	0.3483	0.1808	0.0523
3	0.9999	0.9984	0.9919	0.9744	0.9375	0.8704	0.7599	0.5904	0.3439
4	1.0000	1.0000	1.0000	1.0000	1.0000	1.0000	1.0000	1.0000	1.0000

Table B. Cumulative binomial probabilities (*continues*)

$n = 5$

					p				
k	0.1	0.2	03	0.4	0.5	0.6	0.7	0.8	0.9
0	0.5905	0.3277	0.1681	0.0778	0.0312	0.0102	0.0024	0.0003	0.0000
1	0.9185	0.7373	0.5282	0.3370	0.1875	0.0870	0.0308	0.0067	0.0005
2	0.9914	0.9421	0.8369	0.6826	0.5000	0.3174	0.1631	0.0579	0.0086
3	0.9995	0.9933	0.9692	0.9130	0.8125	0.6630	0.4718	0.2627	0.0815
4	1.0000	0.9997	0.9976	0.9898	0.9688	0.9222	0.8319	0.6723	0.4095
5	1.0000	1.0000	1.0000	1.0000	1.0000	1.0000	1.0000	1.0000	1.0000

$n = 6$

					p				
k	0.1	0.2	03	0.4	0.5	0.6	0.7	0.8	0.9
0	0.5314	0.2621	0.1176	0.0467	0.0156	0.0041	0.0007	0.0000	0.0000
1	0.8857	0.6554	0.4202	0.2333	0.1094	0.0410	0.0109	0.0016	0.0000
2	0.9841	0.9011	0.7443	0.5443	0.3438	0.1792	0.0705	0.0170	0.0013
3	0.9987	0.9830	0.9295	0.8208	0.6562	0.4557	0.2557	0.0989	0.0159
4	0.9999	0.9984	0.9891	0.9590	0.8906	0.7667	0.5798	0.3446	0.1143
5	1.0000	0.9999	0.9993	0.9959	0.9844	0.9533	0.8824	0.7379	0.4686
6	1.0000	1.0000	1.0000	1.0000	1.0000	1.0000	1.0000	1.0000	1.0000

$n = 7$

					p				
k	0.1	0.2	03	0.4	0.5	0.6	0.7	0.8	0.9
0	0.4783	0.2097	0.0824	0.0280	0.0078	0.0016	0.0002	0.0000	0.0000
1	0.8503	0.5767	0.3294	0.1586	0.0625	0.0188	0.0038	0.0004	0.0000
2	0.9743	0.8520	0.6471	0.4199	0.2266	0.0963	0.0288	0.0047	0.0002
3	0.9973	0.9667	0.8740	0.7102	0.5000	0.2898	0.1260	0.0333	0.0027
4	0.9998	0.9953	0.9712	0.9037	0.7734	0.5801	0.3529	0.1480	0.0257
5	1.0000	0.9996	0.9962	0.9812	0.9375	0.8414	0.6706	0.4233	0.1497
6	1.0000	1.0000	0.9998	0.9984	0.9922	0.9720	0.9176	0.7903	0.5217
7	1.0000	1.0000	1.0000	1.0000	1.0000	1.0000	1.0000	1.0000	1.0000

Table B. Cumulative binomial probabilities (*continues*)

$n = 8$

					p				
k	0.1	0.2	03	0.4	0.5	0.6	0.7	0.8	0.9
0	0.4305	0.1678	0.0576	0.0168	0.0039	0.0007	0.0000	0.0000	0.0000
1	0.8131	0.5033	0.2553	0.1064	0.0352	0.0085	0.0013	0.0000	0.0000
2	0.9619	0.7969	0.5518	0.3154	0.1445	0.0498	0.0113	0.0012	0.0000
3	0.9950	0.9437	0.8059	0.5941	0.3633	0.1737	0.0580	0.0104	0.0004
4	0.9996	0.9896	0.9420	0.8263	0.6367	0.4059	0.1941	0.0563	0.0050
5	1.0000	0.9988	0.9887	0.9502	0.8555	0.6846	0.4482	0.2031	0.0381
6	1.0000	0.9999	0.9987	0.9915	0.9648	0.8936	0.7447	0.4967	0.1869
7	1.0000	1.0000	0.9999	0.9993	0.9961	0.9832	0.9424	0.8322	0.5695
8	1.0000	1.0000	1.0000	1.0000	1.0000	1.0000	1.0000	1.0000	1.0000

$n = 9$

					p				
k	0.1	0.2	03	0.4	0.5	0.6	0.7	0.8	0.9
0	0.3874	0.1342	0.0404	0.0101	0.0020	0.0003	0.0000	0.0000	0.0000
1	0.7748	0.4362	0.1960	0.0705	0.0195	0.0038	0.0004	0.0000	0.0000
2	0.9470	0.7382	0.4628	0.2318	0.0898	0.0250	0.0043	0.0003	0.0000
3	0.9917	0.9144	0.7297	0.4826	0.2539	0.0994	0.0253	0.0031	0.0000
4	0.9991	0.9804	0.9012	0.7334	0.5000	0.2666	0.0988	0.0196	0.0009
5	0.9999	0.9969	0.9747	0.9006	0.7461	0.5174	0.2703	0.0856	0.0083
6	1.0000	0.9997	0.9957	0.9750	0.9102	0.7682	0.5372	0.2618	0.0530
7	1.0000	1.0000	0.9996	0.9962	0.9805	0.9295	0.8010	0.5638	0.2252
8	1.0000	1.0000	1.0000	0.9997	0.9980	0.9899	0.9596	0.8658	0.6126
9	1.0000	1.0000	1.0000	1.0000	1.0000	1.0000	1.0000	1.0000	1.0000

$n = 10$

					p				
k	0.1	0.2	03	0.4	0.5	0.6	0.7	0.8	0.9
0	0.3487	0.1074	0.0282	0.0060	0.0010	0.0001	0.0000	0.0000	0.0000
1	0.7361	0.3758	0.1493	0.0464	0.0107	0.0017	0.0000	0.0000	0.0000
2	0.9298	0.6778	0.3828	0.1673	0.0547	0.0123	0.0016	0.0000	0.0000
3	0.9872	0.8791	0.6496	0.3823	0.1719	0.0548	0.0106	0.0009	0.0000
4	0.9984	0.9672	0.8497	0.6331	0.3770	0.1662	0.0473	0.0064	0.0001
5	0.9999	0.9936	0.9527	0.8338	0.6230	0.3669	0.1503	0.0328	0.0016
6	1.0000	0.9991	0.9894	0.9452	0.8281	0.6177	0.3504	0.1209	0.0128
7	1.0000	0.9999	0.9984	0.9877	0.9453	0.8327	0.6172	0.3222	0.0702
8	1.0000	1.0000	0.9999	0.9983	0.9893	0.9536	0.8507	0.6242	0.2639
9	1.0000	1.0000	1.0000	0.9999	0.9990	0.9940	0.9718	0.8926	0.6513
10	1.0000	1.0000	1.0000	1.0000	1.0000	1.0000	1.0000	1.0000	1.0000

Table B. Cumulative binomial probabilities (*continues*)

n = 11

					p				
k	0.1	0.2	03	0.4	0.5	0.6	0.7	0.8	0.9
0	0.3140	0.0859	0.0198	0.0036	0.0005	0.0000	0.0000	0.0000	0.0000
1	0.6970	0.3220	0.1130	0.0302	0.0059	0.0007	0.0000	0.0000	0.0000
2	0.9100	0.6170	0.3130	0.1190	0.0327	0.0059	0.0006	0.0000	0.0000
3	0.9810	0.8390	0.5700	0.2960	0.1130	0.0293	0.0043	0.0002	0.0000
4	0.9970	0.9500	0.7900	0.5330	0.2740	0.0994	0.0216	0.0020	0.0000
5	1.0000	0.9880	0.9220	0.7530	0.5000	0.2470	0.0782	0.0117	0.0003
6	1.0000	0.9980	0.9780	0.9010	0.7260	0.4670	0.2100	0.0504	0.0028
7	1.0000	1.0000	0.9960	0.9710	0.8870	0.7040	0.4300	0.1610	0.0185
8	1.0000	1.0000	0.9990	0.9940	0.9670	0.8810	0.6870	0.3830	0.0896
9	1.0000	1.0000	1.0000	0.9990	0.9940	0.9700	0.8870	0.6780	0.3030
10	1.0000	1.0000	1.0000	1.0000	1.0000	0.9960	0.9800	0.9140	0.6860
11	1.0000	1.0000	1.0000	1.0000	1.0000	1.0000	1.0000	1.0000	1.0000

n = 12

					p				
k	0.1	0.2	03	0.4	0.5	0.6	0.7	0.8	0.9
0	0.2824	0.0687	0.0138	0.0022	0.0002	0.0000	0.0000	0.0000	0.0000
1	0.6590	0.2749	0.0850	0.0196	0.0032	0.0003	0.0000	0.0000	0.0000
2	0.8891	0.5583	0.2528	0.0834	0.0193	0.0028	0.0002	0.0000	0.0000
3	0.9744	0.7946	0.4925	0.2253	0.0730	0.0153	0.0017	0.0000	0.0000
4	0.9957	0.9274	0.7237	0.4382	0.1938	0.0573	0.0095	0.0006	0.0000
5	0.9995	0.9806	0.8822	0.6652	0.3872	0.1582	0.0386	0.0039	0.0000
6	0.9999	0.9961	0.9614	0.8418	0.6128	0.3348	0.1178	0.0194	0.0005
7	1.0000	0.9994	0.9905	0.9427	0.8062	0.5618	0.2763	0.0726	0.0043
8	1.0000	0.9999	0.9983	0.9847	0.9270	0.7747	0.5075	0.2054	0.0256
9	1.0000	1.0000	0.9998	0.9972	0.9807	0.9166	0.7472	0.4417	0.1109
10	1.0000	1.0000	1.0000	0.9997	0.9968	0.9804	0.9150	0.7251	0.3410
11	1.0000	1.0000	1.0000	1.0000	0.9998	0.9978	0.9862	0.9313	0.7176
12	1.0000	1.0000	1.0000	1.0000	1.0000	1.0000	1.0000	1.0000	1.0000

n = 15

					p				
k	0.1	0.2	03	0.4	0.5	0.6	0.7	0.8	0.9
0	0.2059	0.0352	0.0047	0.0005	0.0000	0.0000	0.0000	0.0000	0.0000
1	0.5490	0.1671	0.0353	0.0052	0.0005	0.0000	0.0000	0.0000	0.0000
2	0.8159	0.3980	0.1268	0.0271	0.0037	0.0003	0.0000	0.0000	0.0000

Table B.   Cumulative binomial probabilities (*continues*)

					p				
k	0.1	0.2	03	0.4	0.5	0.6	0.7	0.8	0.9
3	0.9444	0.6482	0.2969	0.0905	0.0176	0.0019	0.0000	0.0000	0.0000
4	0.9873	0.8358	0.5155	0.2173	0.0592	0.0093	0.0007	0.0000	0.0000
5	0.9978	0.9389	0.7216	0.4032	0.1509	0.0338	0.0037	0.0001	0.0000
6	0.9997	0.9819	0.8689	0.6098	0.3036	0.0950	0.0152	0.0008	0.0000
7	1.0000	0.9958	0.9500	0.7869	0.5000	0.2131	0.0500	0.0042	0.0000
8	1.0000	0.9992	0.9848	0.9050	0.6964	0.3902	0.1311	0.0181	0.0003
9	1.0000	0.9999	0.9963	0.9662	0.8491	0.5968	0.2784	0.0611	0.0022
10	1.0000	1.0000	0.9993	0.9907	0.9408	0.7827	0.4845	0.1642	0.0127
11	1.0000	1.0000	0.9999	0.9981	0.9824	0.9095	0.7031	0.3518	0.0556
12	1.0000	1.0000	1.0000	0.9997	0.9963	0.9729	0.8732	0.6020	0.1841
13	1.0000	1.0000	1.0000	1.0000	0.9995	0.9948	0.9647	0.8329	0.4510
14	1.0000	1.0000	1.0000	1.0000	1.0000	0.9995	0.9953	0.9648	0.7941
15	1.0000	1.0000	1.0000	1.0000	1.0000	1.0000	1.0000	1.0000	1.0000

$n = 20$

					p				
k	0.1	0.2	03	0.4	0.5	0.6	0.7	0.8	0.9
0	0.1216	0.0115	0.0008	0.0000	0.0000	0.0000	0.0000	0.0000	0.0000
1	0.3917	0.0692	0.0076	0.0005	0.0000	0.0000	0.0000	0.0000	0.0000
2	0.6769	0.2061	0.0355	0.0036	0.0000	0.0000	0.0000	0.0000	0.0000
3	0.8670	0.4114	0.1071	0.0160	0.0013	0.0000	0.0000	0.0000	0.0000
4	0.9568	0.6296	0.2375	0.0510	0.0059	0.0003	0.0000	0.0000	0.0000
5	0.9887	0.8042	0.4164	0.1256	0.0207	0.0016	0.0000	0.0000	0.0000
6	0.9976	0.9133	0.6080	0.2500	0.0577	0.0065	0.0003	0.0000	0.0000
7	0.9996	0.9679	0.7723	0.4159	0.1316	0.0210	0.0013	0.0000	0.0000
8	0.9999	0.9900	0.8867	0.5956	0.2517	0.0565	0.0051	0.0001	0.0000
9	1.0000	0.9974	0.9520	0.7553	0.4119	0.1275	0.0171	0.0006	0.0000
10	1.0000	0.9994	0.9829	0.8725	0.5881	0.2447	0.0480	0.0026	0.0000
11	1.0000	0.9999	0.9949	0.9435	0.7483	0.4044	0.1133	0.0100	0.0000
12	1.0000	1.0000	0.9987	0.9790	0.8684	0.5841	0.2277	0.0321	0.0004
13	1.0000	1.0000	0.9997	0.9935	0.9423	0.7500	0.3920	0.0867	0.0024
14	1.0000	1.0000	1.0000	0.9984	0.9793	0.8744	0.5836	0.1958	0.0113
15	1.0000	1.0000	1.0000	0.9997	0.9941	0.9490	0.7625	0.3704	0.0432
16	1.0000	1.0000	1.0000	1.0000	0.9987	0.9840	0.8929	0.5886	0.1330
17	1.0000	1.0000	1.0000	1.0000	0.9998	0.9964	0.9645	0.7939	0.3231
18	1.0000	1.0000	1.0000	1.0000	1.0000	0.9995	0.9924	0.9308	0.6083
19	1.0000	1.0000	1.0000	1.0000	1.0000	1.0000	0.9992	0.9885	0.8784
20	1.0000	1.0000	1.0000	1.0000	1.0000	1.0000	1.0000	1.0000	1.0000

Table B.   Cumulative binomial probabilities

References and Further Readings

Chapter 1

Bennet, Deborah, *Randomness*. Cambridge, MA: Harvard University Press, 1998.

Bernstein, Peter, *Against the Gods: The Remarkable Story of Risk*. New York: John Wiley & Sons, 1996.

David, Florence Nightingale, *Games, Gods, and Gambling*. New York: Dover Publications, 1998.

Everitt, Brian S., *Chance Rules: An Informal Guide to Probability, Risk, and Statistics*. New York: Springer-Verlag, 1999.

Feller, William, *An Introduction to Probability Theory and its Applications*, 2 vols., New York: John Wiley & Sons, 1968.

Galilei, Galileo, Sopra le Scoperte dei Dadi, *Opere, Firenze, Barbera*, 1898; 8, pp. 591–594.

Hald, Anders, *A History of Probability and Statistics and Their Applications before 1750*. New York: John Wiley & Sons, 1990.

———, *A History of Mathematical Statistics From 1750 to 1930*. New York: John Wiley & Sons, 1998.

Keynes, John M., *A Treatise on Probability*. New York: Harper and Row, 1992.

McGervey, John D., *Probabilities in Everyday Life*. New York: Ivy Books, 1992.

Packel, Edward, *The Mathematics of Games and Gambling*. Washington, DC: Mathematical Association of America, 1981.

Ross, Ken, *A Mathematician at the Ballpark: Odds and Probabilities for Baseball Fans*. New York: Pearson Education, 2004.

Chapter 2

Albert, James, *Teaching Statistics Using Baseball*. Washington, DC: Mathematical Association of America, 2003.

Gould, Stephen J., *The Mismeasure of Man*. New York: W. W. Norton, 1996.

Hald, Anders, *A History of Probability and Statistics and Their Applications before 1750*. New York: John Wiley & Sons, 1990.

————, *A History of Mathematical Statistics From 1750 to 1930*. New York: John Wiley & Sons, 1998.

Jacoby, Russell, and Glauberman, Naomi (eds), *The Bell Curve Debate: History, Documents, Opinions*. New York: Times Books, 1995.

Porter, T.M., *The Rise of Statistical Thinking 1820–1900*. Princeton: Princeton University Press, 1986.

Quetelet, Adolphe, *Letters Addressed to H. R. H. the Grand Duke of Saxe Coburg and Gotha, on the Theory of Probabilities as Applied to the Moral and Political Sciences*, trans. O. G. Downes. London: Layton, 1849.

Stigler, Stephen M., *The History of Statistics: The Measurement of Uncertainty Before 1900*. Cambridge: The Belknap Press of Harvard University Press, 1986.

————, *Statistics on the Table*. Cambridge: Harvard University Press, 1999.

Walpole Ronald E., Myers, Raymond H., Myers, Sharon L., Ye, Keying, and Yee, Keying, *Probability and Statistics for Engineers and Scientists*, 7th edition. New Jersey: Prentice Hall, 2002.

Chapter 3

Arrow, Kenneth. 1963. *Social Choice and Individual Values*, 2nd ed. New Haven: Yale University Press.

Barry, Brian W. and Russell Hardin. 1982. *Rational Man and Irrational Society? An Introduction and Sourcebook*. Beverly Hills: Sage Publications.

Bauer, Alissa. December 8, 2004. "Rank System Strips Teams of Postseason Opportunities." *University Daily Kansan*, p. 2.

Black, Duncan. 1958. *The Theory of Committees and Elections*. Cambridge: Cambridge University Press.

Borda, Jean-Charles de. 1781. *Mémoire sur les Élections au Scrutin'*. Paris: Histoire del l'Académie Royale des Sciences.

Condorcet, Marquis de. 1785. *Éssai sur L'application L'analyse à la probabilité dés Décisions Rendues à la Pluralité des Voix* (trans. *Essay on the Application of Mathematics to the Theory of Decision-Making*). Paris: L'Imprimerie Royale.

Devlin, Keith. 2004. "Election Math." *Devlin's Angle, MAA Online. http://www.maa.org/devlin/devlin_114_04.html.*

Fishburn, Peter C. 1974. "Paradoxes of Voting." *American Political Science Review* 68(2):537–556.

Harsanyi, John C. 1955. "Cardinal Welfare, Individualistic Ethics, and Interpersonal Comparisons of Utility." *Journal of Political Economy* 61:309–321.

Hilburn, Robert. February 28, 2002. "It's Time to Nominate a New System." *Los Angeles Times*.

Johnson, Paul E. 1998. *Social Choice: Theory and Research*. Thousand Oaks, CA: Sage.

Jones, Bradford, Benjamin Radcliffe, Charles Taber, and Richard Timpone. 1995. "Condorcet Winners and the Paradox of Voting." *American Political Science Review* 89(1):137–144.

Mackenzie, Dana. 2000a. "Making Sense Out of Consensus: How Did That Guy Win?" *SIAM News* 33.

Mackenzie, Dana. 2000b. "May the Best Man Lose." *Discover* 21(11):85–91.

May, Kenneth O. 1952. "A Set of Independent Necessary and Sufficient Conditions for Simple Majority Decision." *Econometrica* 20:680–784.

Nash, John F. 1950. "The Bargaining Problem." *Econometrica* 18:155–162.

Nelson, Randy A., Michael R. Donihue, Donald M. Waldman, and Calbraith Wheaton. 2001. "What's An Oscar Worth?" *Economic Inquiry* 39(1):1–16.

Novaselic, Krist. 2004. "Instant Runoff Voting Used for First Time in San Francisco Elections." *http://www.fixour.us/editorials/11.08.04.html.*

Novoselic, Krist. 2004. *Of Grunge & Government: Let's Fix This Broken Democracy!* New York: Akashic Books.

Reilly, Benjamin. 2002. "Social Choice in the South Seas: Electoral Innovation and the Borda Count in the Pacific Island Countries." *International Political Science Review* 23(4):355–372.

Riker, William H. 1982. *Liberalism Against Populism: A Confrontation Between the Theory of Democracy and the Theory of Social Choice.* San Francisco, CA: W. H. Freeman.

Riker, William H. 1986. *The Art of Political Manipulation.* New Haven, CT: Yale University Press.

Saari, Donald G. 1994. *Geometry of Voting.* Berlin: Springer-Verlag.

Schulze, Markus. 2003. "A New Monotonic and Clone-Independent Single-Winner Election Method." *Voting Matters* 17:9–19.

Woodling, Chuck. December 24, 2004. "Maybe It's Time To Do Away with AP Poll." *Lawrence Journal World*, p. 3C.

Chapter 4

Coleman, A., *Game Theory and Experimental Games*, Pergamon Press, Oxford, 1982.

Davenport, W. C., Jamaican Fishing Village, *Papers in Carribean Anthropology*, **59**, (I. Rouse ed.) Yale University Press, 1960.

Davis, M. D., *Game Theory*, Basic Books, New York, 1983.

Dixit, A. and Nalebuff B., *Thinking Strategically*, W. W. Norton & Company, New York, 1991.

Dresher, M., *The Mathematics of Games and Strategy, Theory and Applications*, Dover, New York, 1981.

Gale, D., *The Theory of Linear Economic Models*, McGraw Hill, New York, 1960.

Haywood O. G. Junior, Military Decision and Game Theory. *J. of the Operations Research Society of America*, **2**(1954), 365–385.

Heckathorn, D. D., The Dynamics and Dilemmas of Collective action, *American Sociological Review*, **61**(1996) No. 2, 250–277.

Jones, A. J., *Game Theory: Mathematical Models of Conflict*, Halstead Press, Chichester, England.

Luce, R. D., and Raiffa, H., *Games and Decision: Introduction and Critical Survey*, John Wiley and sons, London 1957.

Maynard Smith, J. and Parker, G. A., The Logic of Animal Conflict, *Nature*, **246**(1973), 15–18.

Maynard Smith, J., *Evolution and the Theory of Games*, Cambridge University Press, London 1982.

Montgomery J. D., Equilibrium Wage Dispersion and Interindustry Wage Differentials, *Quarterly Journal of Economics* 1991, 163–179.

Morris, P., *Introduction to Game Theory*, Springer-Verlag, New York, 1994.

Nash, J., Non-cooperative games, *Annals of Mathematics*, **54**(1950), 286–295.

Packel E., *The Mathematics of Games and Gambling*, The Mathematical Association of America, Washington D. C., 1981.

Poundstone, W., *Prisoner's Dilemma*, Doubleday, New York, 1992.

Rapoport A., Guyer, M. J., and Gordon, D. G., *The 2 x 2 Game*, The University of Michigan Press, Ann Arbor, 1976.

Sinervo, B., and Lively, C. M., The Rock-Paper-Scissors game and the evolution of alternative male strategies, *Nature*, **380**(1996), 240–243.

Straffin P. D., *Game Theory and Strategy*, The Mathematical Association of America, Washington D. C., 1993.

Thomas L. C., *Games, Theory and Applications*, Ellis Horwood Limited, Chichester England, 1984.

Ventsel, Y. S., *Elements of Game Theory*, Mir Publishers, Moscow, 1980.

von Neumann, J., Zur Theorie der Geselltschaftsspiele, *Mathematische Annalen*, **100**(1928), 295–300.

von Neumann, J. and Morgenstern O., *Theory of Games and Economic Behavior*, Princeton University Press, Princeton 1947.

Williams, J. D., *The Compleat Strategyst*, Dover, New York, 1986.

Zagare, F. C., *Game Theory: Concepts and Applications*, Sage Publications, Beverly Hills, 1984.

Chapter 5

Dorfman, Robert, Samuelson, Paul A., and Solow, Robert M., *Linear Programming and Economic Analysis*. New York: Dover Publications, 1987.

Gass, Saul I., *An Illustrated Guide to Linear Programming*. New York: McGraw-Hill, 1970.

———, *Decision Making, Models, and Algorithms*. Florida: Krieger, 1991.

Kolman, Bernard and Beck, Robert E., *Elementary Linear Programming with Applications*, 2nd edition. New York, Academic Press, 1995.

Solutions to Selected Exercises

Exercises 1.1

1. a) $10/36 = 27.8\%$ b) $16/36 = 44.4\%$ c) $8/36 = 22.2\%$
   d) $24/36 = 66.7\%$ e) $9/36 = 25\%$ f) $9/36 = 25\%$
   g) $10/36 = 27.8\%$ h) $11/36 = 30.6\%$ i) $35/36 = 97.2\%$

3. $(1, 1)$ $(1, 2)$ $(1, 3)$ $(1, 4)$
   $(2, 1)$ $(2, 2)$ $(2, 3)$ $(2, 4)$
   $(3, 1)$ $(3, 2)$ $(3, 3)$ $(3, 4)$
   $(4, 1)$ $(4, 2)$ $(4, 3)$ $(4, 4)$

5. a) $7/24 = 29.2\%$ b) $11/24 = 45.8\%$ c) $6/24 = 25\%$
   d) $17/24 = 70.8\%$ e) $6/24 = 25\%$ f) $6/24 = 25\%$
   g) $8/24 = 33.3\%$ h) $9/24 = 37.5\%$ i) $23/24 = 95.8\%$

7. a) $22/144 = 15.5\%$ b) $34/144 = 23.6\%$ c) $20/144 = 13.9\%$
   d) $54/144 = 37.5\%$ e) $36/144 = 25\%$ f) $36/144 = 25\%$
   g) $22/144 = 15.5\%$ h) $23/144 = 16\%$ i) $143/144 = 99.3\%$

9. a) $2/144 = 1.4\%$ b) $3/144 = 2.1\%$ c) $4/144 = 2.8\%$
   d) $5/144 = 3.5\%$ e) $6/144 = 4.2\%$ f) $7/144 = 4.9\%$
   g) $8/144 = 5.6\%$ h) $9/144 = 6.2\%$ i) $10/144 = 6.9\%$
   j) $11/144 = 7.6\%$ k) $12/144 = 8.3\%$ l) $11/144 = 7.6\%$
   m) $10/144 = 6.9\%$ n) $9/144 = 6.2\%$

Exercises 1.2

1. a) 1/216 = 0.5%   b) 3/216 = 1.4%   c) 6/216 = 2.8%
   d) 10/216 = 4.6%   e) 15/216 = 6.9%   f) 21/216 = 9.7%
   g) 27/216 = 12.5%   h) 25/216 = 11.6%   i) 21/216 = 9.7%
   j) 15/216 = 6.9%   k) 10/216 = 4.6%   l) 6/216 = 2.8%
   m) 3/216 = 1.4%   n) 1/216 = 0.5%

3. a) 1/64 = 1.5%   b) 3/64 = 4.7%   c) 6/64 = 9.4%
   d) 10/64 = 15.6%   e) 12/64 = 18.8%   f) 12/64 = 18.8%
   g) 10/64 = 15.6%

5. a) 1 + 1 + 1 + 1   b) 2 + 1 + 1 + 1
   c) 3 + 1 + 1 + 1, 2 + 2 + 1 + 1
   d) 4 + 1 + 1 + 1, 3 + 2 + 1 + 1, 2 + 2 + 2 + 1
   e) 5 + 1 + 1 + 1, 4 + 2 + 1 + 1, 3 + 3 + 1 + 1, 3 + 2 + 2 + 1, 2 + 2 + 2 + 2
   f) 6 + 1 + 1 + 1, 5 + 2 + 1 + 1, 4 + 3 + 1 + 1, 4 + 2 + 2 + 1, 3 + 3 + 2 + 1,
      3 + 2 + 2 + 2
   g) 7 + 1 + 1 + 1, 6 + 2 + 1 + 1, 5 + 3 + 1 + 1, 5 + 2 + 2 + 1, 4 + 4 + 1 + 1,
      4 + 3 + 2 + 1, 4 + 2 + 2 + 2, 3 + 3 + 3 + 1, 3 + 3 + 2 + 2

7. 3321, 3312, 3231, 3213, 3123, 3132, 2313, 2331, 2133, 1332, 1323, 1233

9. 2111, 1211, 1121, 1112

11. a) 1/256 = 0.4%   b) 4/256 = 1.6%   c) 10/256 = 3.9%
    d) 20/256 = 7.8%   e) 31/256 = 12.11% f) 40/256 = 15.6%
    g) 44/256 = 17.2%

13. a) 1/4096 = 0.02%   b) 4/4096 = 0.1%   c) 10/4096 = 0.2%
    d) 20/4096 = 0.5%   e) 35/4096 = 0.8%   f) 56/4096 = 1.1%
    g) 84/4096 = 2.1%

Exercises 1.3

1. a) 40.2%   b) 40.2%   c) 16.1%   d) 0%
3. a) 0%   b) 0.1%   c) 31.1%   d) 13.3%
5. a) 13.2%   b) 32.9%   c) 0.4%

7. a) 37.7%   b) 39.9%   c) 17.6%
9. a) 86.8%   b) 25.2%   c) 12%   d) 73.6%
11. a) 49.7%   b) 79.7%   c) 29.4%   d) 48.6%
13. a) 34.1%   b) 88.9%   c) 23%   d) 71.4%

Exercises 1.4

1. a) 8.9%   b) 93.1%   c) 5.6%   d) 86.2%
3. a) 69.8%   b) 38.3%   c) 73.8%   d) 8.1%
5. a) 77.8%   b) 24.4%   c) 2.2%   d) 76.5%
7. a) 18%   b) 89.9%   c) 14.5%   d) 17.8%
9. a) 4.6%   b) 96.8%   c) 7%   d) 93.7%
11. a) 99.7%   b) 0.3%   c) 0.08%   d) 97.1%

Exercises 2.1

1. 1   3. 3.53   5. 0.47

7.

Index	Data	Frequency	Relative F.
1	60	5	12.5%
2	61	3	7.5%
3	62	3	7.5%
4	63	4	10 %
5	64	5	12.5%
6	65	3	7.5%
7	66	2	5 %
8	67	6	15 %
9	68	4	10 %
10	69	5	12.5%
		40	100%

Mean = 64.7, standard deviation = 3.0.

9.

Index	Group	Frequency	Relative F.	Mark
1	5–9	1	2%	7
2	10–14	5	10%	12
3	15–19	8	16%	17
4	20–24	12	24%	22
5	25–29	13	26%	27
6	30–34	6	12%	32
7	35–39	5	10%	37
		50	100%	

Mean = 23.9, standard deviation = 7.5.

11.

Index	Group	Frequency	Relative F.	Mark
1	30–36	9	4.5%	33
2	37–43	73	36.5%	40
3	44–50	96	48 %	47
4	51–57	9	4.5%	54
5	58–64	7	3.5%	61
6	65–71	6	3 %	68
		200	100%	

Mean = 45.25, standard deviation = 6.8.

13.

Index	Group	Frequency	Relative F.	Mark
1	16–18	111	1.9%	17
2	19–21	1380	23.4%	20
3	22–24	1158	19.7%	23
4	25–27	928	15.7%	26
5	28–30	847	14.4%	29
6	31–33	591	10.0%	32
7	34–36	878	14.9%	35

Mean = 26.2, standard deviation = 5.35

15. Mean = 2; $(cc) = 4(1 - 2) + 4(3 - 2) = -4 + 4 = 0.$

Exercises 2.2

1. a) Median = 3, average = 4

   b) $|1-3| + |2-3| + |3-3| + |4-3| + |10-3| = 11$

   $|1-4| + |2-4| + |3-4| + |4-4| + |10-4| = 12$

   c) $|1-3|^2 + |2-3|^2 + |3-3|^2 + |4-3|^2 + |10-3|^2 = 55$

   $|1-4|^2 + |2-4|^2 + |3-4|^2 + |4-4|^2 + |10-4|^2 = 50$

3. a) Median = 5.3, average = 5.4   b) 1.1, 1.2   c) 0.55, 0.50

5. a) 9.1, 9.14   b) 0.2, 0.24   c) 0.02, 0.012

7. a) 8.6, 8.6   b) 0.6, 0.6   c) 0.06, 0.06

Exercises 2.3

1. a) $y = -x + 2$ b) (0, 0)   c) 0

   e) $y = -x + 2$, (0, 0), 0 = 0.

3. a) $y = x$ b) (0, 0, 0)   c) 3

   e) $y = x$, (0, 0, 0), 0 = 0

5. a) $y = 7/3$ b) (–1/3, 2/3, –1/3)   c) 7/3

   e) $y = 2$, (0, 1, 0), 2/3 < 1

7. a) $y = -1.06x + 36.29$ b) (0.31, –0.39, 0.21)   c) 15.09

   e) $y = -1.06x + 36.7$, (0, –2/3, 0), 0.292 < 0.444.

9. $y = -316.3x + 6437.3$, 3274.4

11. $y = 1.58x + 133.51$, 212.42

Exercises 2.4

1. a) 70.3%   b) 29.6   c) 88.4%

   d) 89.4%   e) 72.1%   f) 15.3%

3. a) 78   b) 81   c) 65

   d) 67   e) 74   f) 70

5. a) 11,949   b) 2,536   c) 3084   d)11,575

   e) 3423

   f) Either 0 or 1271, depending on the interpretation of "exactly."

   g) 6847

7. a) 4.8%   b) 90.4%   c) 4.8%

   d) 70.65%   e) 9.6%   f) 1.2%

9. a) 62.5%   b) 99.4%   c) 30.2%   d) 30.9%

   e) 30.2%   f) 93.3%   g) 30.2%   h) 18.4%

11. a) 15.9%   b) 84%   c) 0.1%

    d) 49.5%   e) 68.3%   f) 89%

13. a) 44%   b) 84.1%   c) 14.6%   d) 59.9%

    e) 77.5%   f) 15.9%   g) 45.3%   h) 26.1%

15. a) [25, 29.8], [22.6, 32.2], [20.2, 34.6]

    b) 68.3%, 95.4%, 99.7%

Exercises 2.5

1. a) (0.61, 0.82), 0.12, 4145   b) (0.63, 0.80), 0.09, 2401

   c) (0.65, 0.78), 0.075, 1692

3. a) (0,07, 0.17), 0.07, 16577   b) (0.09, 0.16), 0.06, 9604

   c) (0.09, 0.15), 0.05, 6766

5. a) (0.01, 0.12), 0.11, 16577   b) (0.03, 0.11), 0.08, 9604

   c) (0.03, 0.10), 0.07, 6766

7. a) (3.02, 4.18)   b) (3.16, 4.04)   c) (3.23, 3.98)

9. a) (239, 245)   b) (239, 245)   c) (240, 244)

11. p-value = 0.4%. Reject the 60% belief at all levels.

13. p-value = 0.1%. Reject the 90% belief at all levels.

15. p-value = 9.9%. Reject the 75% at level 10% only.

17. p-value = 1.5%. Conclude they are different at levels 5% and 10% only.

19. p-value = 4.6%. Conclude there was a change at the 5% and 10% levels only.

Exercises 2.6

1. b) 2.5, 1.12   d) 2.5, 0.65   f) 2.5, 0.37
3. b) 3.5, 2.06   d) 3.5, 1.19   f) 3.5, 0.69
5. b) 2.25, 0.83   d) 2.25, 0.48   f) 2.25, 0.28
7. b) 2, 0   d) 2, 0   f) 2, 0
9. b) 2.4, 1.02   d) 2.4, 0.62   f) 2.4, 0.42
11. b) 2, 0.63   d) 2, 0.39   f) 2, 0.26
13. b) 3, 0   d) 3, 0   f) 3, 0

Exercises 3.2

1. *steak* $\succ_{Joe}$ *eggplant*
3. Mary's preference violates the principle of transitivity.

Exercises 3.3

1. No
2. This is a well-behaved example because candidate x will win with any procedure.
    a) There is no majority rule winner.
    b) Plurality vote totals: $x = 34$, $y = 16$, $z = 0$. Therefore, x wins.
    c) No
    d) Eliminating z does not affect any plurality votes.
    e) $x \succ_P y$ and $x \succ_P z$
5. Note how the winner changes according to the different methods used.
    a) w
    b) No
    c) $x \succ_P w$
    d) Eliminate y. Plurality totals are $x = 24$, $z = 40$, and $w = 36$. z wins the plurality, but there is no majority winner. Eliminate x, and the final runoff has $w \succ_P z$.

7. Strong monotonicity requires that, if there is a tie, then the change of a single voter's opinion must break the tie. A "hung jury" is the legal equivalent of a tie vote in an election. A jury that is divided 9-3 is "hung," unable to decide. If someone changes, so the total is 10-2, the jury is still unable to decide.

Exercises 3.4

1. Make a table to show that, with 2 candidates, the Borda count and the majority election give the same results.

		Borda Votes		Borda Count
	Group:	1	2	
Number of	Members	81	54	
Alternatives				
Fred		1	0	81
Barney		0	1	54

3. a) Compare each candidate against each of the others:

$x \succ_p y$, $z \succ_p x$, $x \succ_p w$

$y \succ_p w$, $z \succ_p y$

$z \succ_p w$

z is the Condorcet winner.

b) x wins the Borda count. The overall ranking is $x \succ_{BC} z \succ_{BC} y \succ_{BC} w$.

		Borda	Votes					Borda count
	Groups	1	2	3	4	5	6	
Number of	Members	200	150	150	300	100	100	
Alternatives								
w		0	0	0	3	1	0	1000
x		2	1	3	1	2	3	1800
y		1	3	2	0	3	2	1450
z		3	2	1	2	0	1	1750

c) After eliminating w, the Borda ranking is $z \succ_{BC} x \succ_{BC} y$.

		Borda	Votes					Borda count
	Groups	1	2	3	4	5	6	
Number of	Members	200	150	150	300	100	100	
Alternatives								
x		1	0	2	1	1	2	1100
y		0	2	1	0	2	1	750
z		2	1	0	2	0	0	1150

    d) No. Deleting a candidate never affects the Condorcet winner.

5. a) *Pierre*. Note that *Pierre* $\succ_P$ *Paul* with a vote of 41-40.

    b) *Paul*. The Borda totals for (*Pierre, Paul, Jacques*) are $(101, 109, 33)$.

    c) If you change 1 vote, you can make *Paul* the Condorcet winner. On the other hand, if you try to make *Pierre* win the Borda count, you must change 3 ballots. Consider voter group 5, the ones who vote $(0, 2, 1)$. Change 3 of those voters into group 1, so their votes become $(2, 0, 1)$. The Borda totals change to $(107, 103, 33)$. (Note, if you only change 2 voters, the result is a tie).

Exercises 3.5

1.

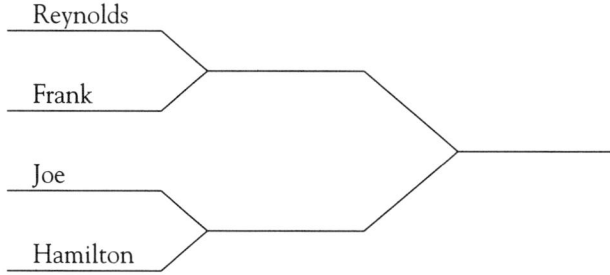

Reynolds

Frank

Joe

Hamilton

3. a) $x \succ_P y$ and $x \succ_P z$

    b) Yes

    c) No, it is impossible.

5 a) Note that x beats y (6-5), y beats z (6-5), and z beats x (8-3).

		Voters					
	Groups	1	2	3	4	5	6
Number of	Members	3	3	2	1	1	1
Ranking							
first		z	y	x	y	y	x
second		x	z	w	z	z	y
third		y	w	z	w	w	z
fourth		w	x	y	x	x	w

b) Our favorite is x. So our tournament would pit y against z in the first round, and y would advance to the final vote, and x would win.

Exercises 3.6

1. The round robin would end up in a tie, with each candidate winning one contest.

3. The Borda totals for (x, y, z) are $(56, 77, 62)$. The aggregated pairwise votes are as follows. Cells represent votes earned for the row candidate against the column candidate.

	x	y	z	Aggregated Pairwise Vote (row sum)
x	.	22	34	56
y	33	.	34	77
z	31	31	.	62

7. Show that if $x \succ_{BC} y \approx_{BC} z$, then the valid pairwise outcomes are:

$$
\begin{aligned}
&1. \quad x \succ_P y, \; x \succ_P z, \; y \approx_P z \\
&2. \quad x \succ_P y, \; x \succ_P z, \; y \succ_P z \\
&3. \quad x \succ_P y, \; z \succ_P x, \; y \succ_P z \\
&4. \quad x \succ_P y, \; x \approx_P z, \; y \succ_P z \\
&5. \quad x \approx_P y, \; x \succ_P z, \; y \succ_P z
\end{aligned}
\tag{1}
$$

Let $q_{x,y}$ to represent the number of voters who prefer x to y.

$$q_{x,y} = |\{i \in N : x \succ_i y\}|.$$

Because there are no indifferent voters, a tie means $q_{x,y} = q_{y,x} = N/2$. Note that

$$q_{x,y} > q_{y,x} \text{ if and only if } q_{x,y} - q_{x,y} > 0 \text{ if and only if } x \succ_P y.$$

The result of the aggregated pairwise vote, which is equivalent to the Borda count, can be expressed as

aggregated pairwise vote for $x : q_{x,y} + q_{x,z}$
aggregated pairwise vote for $y : q_{y,x} + q_{y,z}$
aggregated pairwise vote for $z : q_{z,x} + q_{z,y}$

If the Borda outcome is $x \succ_{BC} y \approx_{BC} z$, *then, by definition*:

$$
\begin{aligned}
q_{x,y} + q_{x,z} &> q_{y,x} + q_{y,z} \\
q_{x,y} + q_{x,z} &> q_{z,x} + q_{z,y} \\
q_{y,z} + q_{y,x} &= q_{z,x} + q_{z,y}
\end{aligned}
\tag{2}
$$

Keep in mind that $q_{y,x} = N - q_{x,y}$, $q_{z,x} = N - q_{x,z}$, and $q_{z,y} = N - q_{y,z}$.

Re-arrange the inequalities so that they have the pairwise majority rule conditions on the left hand side:

$$
\begin{aligned}
q_{x,y} - q_{y,x} &> q_{y,z} - q_{x,z} \\
q_{x,z} - q_{z,x} &> q_{z,y} - q_{x,y} \\
q_{y,z} - q_{z,y} &= q_{z,x} - q_{y,x}
\end{aligned}
\tag{3}
$$

Consider 3 cases: $y \approx_P z$, $y \succ_P z$, and $z \succ_P y$.

Case 1: $y \approx_P z$. We now claim that if this (equivalently, $q_{y,z} = q_{z,y}$) is true, then the previous inequalities imply both $q_{x,y} > q_{y,x}$ and $q_{x,z} > q_{z,x}$. Proceed as follows. Since $q_{z,y} = N - q_{y,z}$, then $q_{y,z} = q_{z,y} = N/2$. The third condition in 3, along with $q_{y,z} = q_{z,y}$, implies $q_{z,x} = q_{y,x}$. That is to say, either x defeats both y and z, and by the same margin, or y and z defeat x, by the same margin. It is obvious that x must defeat both, since it is the winner of the Borda count (and the aggregated pairwise vote).

This means that $x \succ_P z$ and $x \succ_P y$ are implied by the condition $y \approx_P z$, so one word has to be the first one listed in 1. That is to say, in order to be consistent with the Borda outcome $x \succ_{BC} y \approx_{BC} z$, then if y ties z in a plurality election, then x must defeat both y and z (and by the same margin).

Case 2: $y \succ_P z$. Beginning with the information supplied by the Borda outcomes, we can use the identities $q_{x,y} = N - q_{y,x}$, $q_{y,z} = N - q_{z,y}$, and $q_{z,x} = N - q_{x,z}$ to arrive at the following:

$$q_{x,y} - \frac{N}{2} > \frac{1}{2}q_{y,z} - \frac{1}{2}q_{x,z}$$

$$q_{x,z} - \frac{N}{2} > \frac{N}{2} - \frac{1}{2}q_{y,z} - \frac{1}{2}q_{x,y} \qquad (4)$$

$$q_{y,z} = \frac{N}{2} - \frac{1}{2}q_{x,z} - \frac{1}{2}q_{x,y}$$

Insert the value of $q_{y,z}$ implied by the third expression in the second one to obtain

$$q_{x,z} - \frac{N}{2} > \frac{N}{4} + \frac{1}{4}q_{x,z} - \frac{3}{4}q_{x,y}$$

which can be rearranged as:

$$\frac{3}{4}q_{x,z} - \frac{3N}{4} > \frac{3}{4}q_{x,y},$$

or

$$q_{x,z} + q_{x,y} > N \qquad (5)$$

We do not know for sure if x defeats y or x defeats z in pairwise votes, but we can say for sure that x will defeat at least one, or possibly both, of them.

This narrows down the possibilities quite a bit. This case assumes $y \succ_P z$, so 5 implies that the only majority rule outcomes consistent with the Borda result $x \succ_{BC} y I_{BC} z$ are:

$$\begin{array}{ll}
1 & x \succ_P y,\ x \succ_P z,\ y \approx_P z \\
2 & y \succ_P x,\ x \succ_P z,\ y \succ_P z \\
3 & x \succ_P y,\ z \succ_P x,\ y \succ_P z \qquad (6) \\
4 & x \approx_P y,\ z \succ_P z,\ y \succ_P z \\
5 & x \succ_P y,\ x \approx_P z,\ y \succ_P z
\end{array}$$

Possibilities 2 and 4 can be ruled out. Here's why. If $y \succ_P z$, then $z \succ_P x$. To see why, recall $x \succ_{BC} y \approx_{BC} z$, so it must be that

$$q_{y,z} + q_{y,x} = q_{z,x} + q_{z,y}$$

Equivalently,

$$q_{y,z} - q_{z,y} = q_{z,x} - q_{y,x}$$

If $y \succ_P z$, then $q_{y,z} - q_{z,y} > 0$ and so $q_{z,x} - q_{y,x} > 0$ and using the identities like $q_{x,y} + q_{y,x} = N$, we find $q_{x,y} - q_{x,z} > 0$. If $x \succ_P z$, then $q_{x,z} > \dfrac{N}{2}$, and it must be that $q_{x,y} > \dfrac{N}{2}$ (equivalently, $x \succ_P y$). The possibilities 2 and 4 violate that restriction. That leaves possibilities 1, 3, and 5 which correspond to three of the words in the solution.

Case 3: $z \succ_P y$. If we repeat the same exercise as the previous case, we arrive at the same set of possible words, except that the last one "swaps" the positions of y and z.

$$1 \; x \succ_P y, \; x \succ_P z, \; y \succ_P z$$
$$3 \; x \succ_P y, \; z \succ_P x, \; y \succ_P z$$
$$5 \; x \approx_P y, \; x \succ_P z, \; y \succ_P z$$

Notice that two of these solutions are identical to the ones found in the previous case, only the third solution is unique. Collecting up the unique solutions, we have finished our work. All five of the words that are compatible with the Borda outcome $x \succ_{BC} y \approx_{BC} z$ have been found.

Exercises 3.7

1. a) The plurality totals for (x, y, z) are $(22, 12, 31)$, so y is eliminated first. The 12 votes for y are transferred to x.

    b) x wins the instant runoff, $34 - 31$.

    c) Yes, the results agree. x would win a majority/runoff contest.

3. a) The plurality totals for (Hamilton, Joe, Frank, Reynolds) are $(22, 12, 31, 0)$, so Reynolds is eliminated first. Since Reynolds had no first-place votes, deleting him does not change the outcome. Next, Joe is deleted and 12 votes are transferred to Hamilton.

    b) Hamilton wins the instant runoff.

    c) A majority/runoff procedure picks Hamilton and Frank to compete in the runoff election, and Hamilton wins that runoff election.

5.   Note that x is a Condorcet winner, but it is eliminated in the first stage of the instant runoff process. Transfer 10 votes to y, therefore y is the winner.

		Voters		
	Groups:	1	2	3
Number of	Members	10	35	33
Ranking				
first	x	y	z	
second	y	x	x	
third	z	z	y	

Exercises 4.1

1. a) 144°, 108°, 72°, 36°   b) 0   c) 0.5

    d) –0.8   e) 0.1   f) (2, 3)

3. a) 36°, 72°, 108°, 144°   b) 0.5   c) –0.1

    d) –1.3   e) –0.6   f) (1, 3)

5. a) 108°, 252°   b) 83%   c) 85%

    d) 84%   e) 83.8%

Exercises 4.2

1. 2.94   3. 1.8   5. 0.37

7. –3.5   9. 0.12   11. –4

13.

	B-day	No
Flowers	1.5	1
No	–10	0

15.

	Rain	Shine
Umbrellas	250	–150
Glasses	–150	350

Exercises 4.3

1. a) [1, 0], [1, 0]   b) 1, 3   c) (2, 3)   d) All outcomes
    e) None   f) [1, 0], [0, 1], (1, 4)
    g) [1, 0], [1, 0], (2, 3)
3. a) [0, 1], [1, 0]   b) 2, 2   c) (2, 2)   d) (3, 1), (2, 5)
    e) None   f) [0, 1], [0, 1], (2, 5)
    g) [0, 1], [1, 0], (2, 2)
5. a) [1, 0, 0], [1, 0] or [0, 1]   b) 4, 1   c) (5, 4) or (4, 1)
    d) (5, 5)   e) (5, 4), (5,5)
    f) [1, 0, 0], [1, 0], (5, 4) or [0, 0, 1], [0, 1], (5, 5)
7. a) [1, 0, 0], [0, 0, 1]   b) 7, 2   c) (7, 2)   d) All outcomes
    e) (7, 2)   f) [1, 0, 0], [0, 0, 1], (7, 2)
    g) [1, 0, 0], [0, 0, 1], (7, 2)
9. a) [1, 0, 0, 0], [1, 0, 0, 0] or [0, 0, 1, 0]   b) 0, –1
    c) (1, 2) or (1, 1)   d) (0, 3), (3, –1), (2, 2)   e) (2, 2), (2, 2)
    f) [0, 1, 0, 0], [1, 0, 0, 0], (2, 1) or [0, 0, 1, 0], [0, 0, 0, 1], (2, 2) or
       [0, 0, 0, 1], [0, 1, 0, 0], (2, 2)
    g) [0, 0, 0, 1], [0, 1, 0, 0], (2, 2) or [0, 0, 1, 0], [0, 0, 0, 1], (2, 2)
19. Yes

Exercises 4.4

1. (0.86, 1.36)   3. (1.66, 1.88)   5. (0, 1)
7. a) Any strategy   b) [1, 0]
9. a) Any strategy   b) Any strategy
11. a) [0, 0, 1]   b) [0, 1, 0, 0]
13. No   15. Yes   17. Yes
19. Yes   21. No   23. Yes
25. a) 0   b) 1
27. a) 4/11   b) 7/11   29. a) 0.2   b) 0.8   31. a) 2/3   b) 1/3
33. a) 100%   b) None

Exercises 4.5

1. [1/3, 2/3], [2/3, 1/3], 5/3

3. Any strategy, [0, 1], 1

5. [0, 1], [1, 0], 4

7. [1, 0], [1, 0], 1

9. [1, 0], [1, 0], 2

11. [1, 0], any strategy, 1

13. [7/13, 6/13], [4/13, 9/13], 11/13

15. [1, 0], [0, 1], 2

17. [1, 0], [1, 0], −3

19. [1, 0], [1, 0], 0

21. [1, 0], [0, 1], −2

23. [1, 0], [1, 0], 2

25. [3/7, 4/7], [4/7, 3/7], 9/7

27. [0, 1], [0, 1], 0

29. [2/3, 1/3], [2/3, 1/3], −1/3

31. [1, 0], [0, 1, 0], 1

33. [0, 0, 1, 0], [1, 0], 2

35. [0, 1, 0], [0, 1, 0, 0], −2

37. [0, 0, 1, 0], [0, 0, 1, 0], 0

39. [1, 0, 0, 0, 0], [0, 1, 0, 0], 1

41. This game has a saddle point and Frank had better buy the flowers.

43. See Exercise 4.2.15. The oddments are [5:4]. Merrill should invest $161.11 in umbrellas and $88.89 in sunglasses. The value of the game is $72.22.

45. a) Yes   b) Yes   c) No   d) [0.2, 0.8]

    e) No   f) No

    g) [0.6, 0.4], but he probably won't make it.

47. a) Yes   b) Yes   c) No   d) [1,0]

    e) No   f) Yes   g) [0, 1]

Exercises 5.1

See Figures 1 and 2 below.

Figure 1

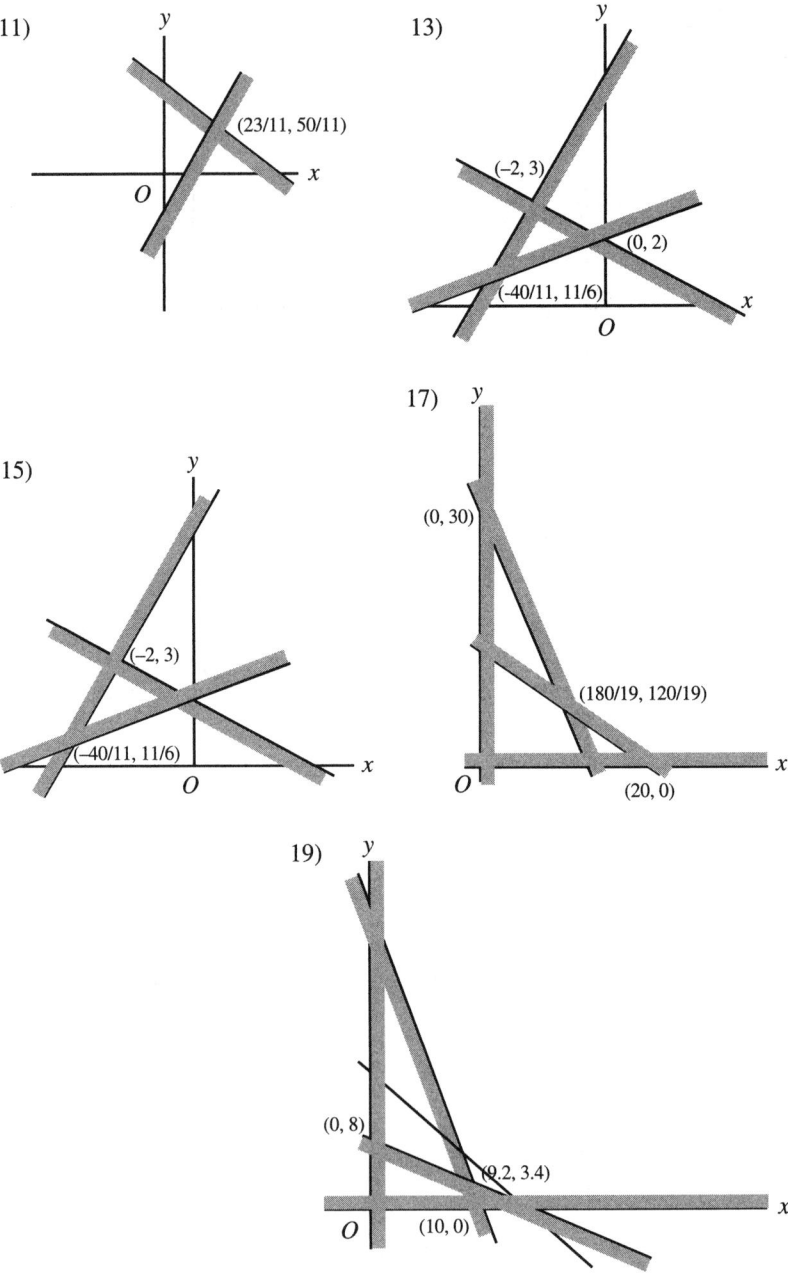

11)

(23/11, 50/11)

13)

(−2, 3)

(0, 2)

(−40/11, 11/6)

15)

(−2, 3)

(−40/11, 11/6)

17)

(0, 30)

(180/19, 120/19)

(20, 0)

19)

(0, 8)

(9.2, 3.4)

(10, 0)

Figure 2

Exercises 5.2

1. 2, 101   3. 2, 92

5. 600 and 2400 lbs in each case.

7. a) 1650   b) 1950   c) 3000

9. a) 4000, 6000   b) 7000, 3000   c) 7000, 3000

11. a) 80,000 and 40,000   b) 80,000 and 40,000   c) 60,000 and 60,000

13. a) 300 and 900   b) 300 and 900   c) 0 and 1200

15. Original question: Detroit 3 weeks and LA 6 weeks at a cost of $3,780,000.

Second question: Detroit 6 weeks, LA 2 weeks at a cost of $4,000,000.

Index

advantage, 12
aggregated pairwise voting, 115–117
anonymity, 90
arithmetic mean, 33
Arrow's theorem, 135
Ars Conjectandi, 15, 72
astragali, 1
atomic error, 56
auxiliary diagram, 151
average, 33

bar graph, 37
Battle of the Bismarck Sea, 191
beatpath method, 119–124
bell-shaped curve, 24
Bernoulli trial, 15
Bernoulli, James, 15, 72
better than, 160
bimodal, 33
binomial
    process, 16
    trial, 15
Bombing Sorties, 141, 197
Book on Games of Chance, 13
Borda count, 96–104
Borel, Emile, 149
Boscovich method, 50

Cardano, Gerolamo, 13
cardinal preferences, 88, 104
Central Limit theorem, 82
central tendency, 33
Ceres, 54
Chicken, 161
choice number, 16

circumference of earth, 50
Cloneproof Schwartz Sequential Dropping
    (CSSD), 119–124
cloneproof, 102
complementary, 5
Condorcet
    criterion, 101–102
    methods, 113
confidence interval, 72
confidence, 72
conflict, 137
conjunction, 5
constraint, 210
correction, 44
Cotes, Roger, 47
counterstrategy, optimal, 171
Cuban missile crisis, 161
cumulative frequency, 60
cumulative relative frequency, 60

De Moivre, Abraham, 72
Dialog Concerning, 55
Doctrine of Chances, 72
dodecahedron, 9
dove, 177

elementary error, 56
error, 44, 55
event, 2
Evolutionary Game, 177
expected payoff, 150, 154
expected value, 145

failure, 15
fitness, 177

France, 50
frequency table, 37

gain, 137
Galilei, Galileo, 10, 43, 55
Galton, Sir Francis, 28
game, 137
    fair, 138
    nonzero-sum, 180
    solution of zero-sum, 188
    solution of, 194
    strictly determined, 191
    zero-sum, 148, 152
Gauss, Carl F., 47, 53
geodesy, 50

hawk, 177
histogram, 37
hypothesis testing, 76

independence from irrelevant alternatives,
    100
independent random events, 145
independent, 5

Jamaican Fishing Village, 198
Job Applicants, 166, 176, 206

Kennedy, John F., 162
Kepler, Johannes, 43, 49
Khrushchov, Nikita, 162

Laplace, Pierre-Simon, 82
least squares method, 47
Legendre, Adrien-Marie, 47, 50, 53
linear programming, 209
Lively, C. M., 139
lizard, 139
loss, 138

majority rule, defined, 90
majority rule, intransitivity of, 109
margin of error, 74
mark, 39

maximin strategy
    mixed, 194
    pure, 158, 188
    value, 158
May's theorem, 90
Maynard Smith, J., 177
mean, 33
    grouped, 40
median, 33, 45
meter, 50
minimax strategy
    mixed, 194
    pure, 189
Minimax theorem, 194
minimax value, pure 189
mixed strategy, 149
mode, 33
model, 51
MOE, 74
monotonicity, 91, 132–133
moral certainty, 72
Morgenstern, Oskar, 137, 150

Nash equilibria of zero-sum games, mixed,
    193
Nash equilibria of zero-sum games, pure,
    190
Nash equilibrium point, pure, 164
Nash equilibrium solution, 171
Nash equilibrium, mixed, 170
Nash, John, 170
neutrality, 90
Newton, Sir Isaac, 47
No Regrets, 164, 170
non-cooperative, 157
nondictatorship, 135
normal curve, 26
normality assumption, 60
normally distributed, 66
null hypothesis, 76

objective function, 210
observation, 44
Occam's razor, 47

octahedron, 7
oddments, 195
odds, 1
optimization, 210, 224
ordinal preferences, 88
outcome, 2, 149, 153

pairwise comparisons, 107–111
paradox, dominated winner, 108–109
paradox, voter's, 110
paradox, winner turns loser, 98
Pareto optimal, 160
Pareto principal, 135
Parker, G. A., 177
participation criterion, 127
partition, 10
payoff pair, 158
payoff, 149, 153
payoff, expected, 154
Penny Matching, 137
Piazzi, Giuseppe, 53
play, 149
plurality rule, 89–93
Polar Star, 55
population, 60
preference relation, binary, 88
Principle of Rationality, 158
Prisoner's Dilemma, 159, 205
probability, 2, 145
p-value, 76

Quetelet, Adolphe, 58
quincunx, 28

random experiment, 2
random variable, 66
range, 35
regression line, 53
relative frequency, 37
River Tale, 196
Rock-Scissors-Paper, 138

Roman Catholic Church, 43
runoff election, 91

saddle point, 191
Schulze method, 119
Schwartz set, 121
significance, 77
simultaneous equations, 222
Sinervo, B., 139
single transferable vote, 130–135
Smith criterion, 114
Smith-IIA criterion, 120
Smith set, 114
spread, 35
standard deviation, 36
    grouped, 40
standard normal curve, 27
strategy pair, mixed, 169
strategy, 137, 150, 153
    mixed, 153, 169
    pure, 153
success, 15
symmetric, 140

tetrahedron, 8
throw, 10
tournament,
    round robin, 113–115
    single elimination, 107
transitivity, 88
Two-Finger Morra, 140

Uta stansburiana, 139

value, 188, 194
value pair
    mixed, 171
    pure, 158
variability, 35
vertex principle, 225
von Neumann, John, 137, 150